VETERINARY INFECTIOUS DISEASES
Identification and Treatment

VETERINARY INFECTIOUS DISEASES
Identification and Treatment

RUCHI SANDILYA
UTKARSH RAJPUT

KNOWLEDGE
BAKERS

KNOWLEDGE BAKERS

ISBN: 978-93-49964-81-5

Published by : **Knowledge Bakers**
Pune, India
Email: knowledgebakers2019@gmail.com

PREFACE

The field of veterinary infectious diseases is a cornerstone of modern veterinary medicine, as it focuses on understanding the myriad pathogens that affect animals and the impact these diseases have on both individual animal health and public health. Infectious diseases in animals pose significant challenges to veterinary practitioners, livestock industries, and wildlife conservationists. They can lead to economic losses, pose threats to public health, and disrupt ecosystems. With the ever-evolving nature of pathogens and the rise of emerging diseases, the need for accurate identification, effective treatment, and preventive strategies has never been more urgent.

This book, Veterinary Infectious Diseases: Identification and Treatment, provides a comprehensive and accessible resource for veterinarians, veterinary students, researchers, and professionals involved in animal health. It is designed to equip readers with the knowledge and tools necessary for the diagnosis, treatment, and management of infectious diseases in animals, ranging from companion animals to livestock, and even wildlife.

The first part of the book introduces the basic principles of veterinary infectious diseases, including the mechanisms of infection, pathogenesis, and the roles of different pathogens such as bacteria, viruses, fungi, and parasites. It also explores the key factors influencing the spread of infectious diseases, including environmental, host-related, and pathogen-related factors. This foundational knowledge sets the stage for understanding how infections occur and how they can be managed effectively.

Following this introduction, the book delves into the specifics of identifying and diagnosing infectious diseases. Detailed chapters on diagnostic techniques, including microbiological methods, serology, molecular biology, and immunohistochemistry, provide readers with a deep understanding of how veterinary professionals identify pathogens and determine the appropriate course of treatment. Each disease is discussed in terms of its clinical signs, diagnostic methods, and treatment options, with a strong focus on practical, evidence-based approaches.

The book also covers the treatment options available for various infectious diseases, emphasizing both pharmacological and non-pharmacological interventions. Chapters on antimicrobial therapy, vaccination strategies, and alternative treatments are included, offering insights into the challenges and advancements in treating infections. The issue of antimicrobial resistance (AMR) is also discussed in detail, stressing the importance of responsible antibiotic use and the role of veterinarians in combating this growing global threat.

In addition to the individual treatment of diseases, Veterinary Infectious Diseases addresses the broader implications of managing infectious outbreaks. It covers topics such as quarantine protocols, biosecurity measures, vaccination programs, and zoonotic diseases—diseases that can be transmitted between animals and humans. The book highlights the importance of an integrated approach to animal health, where veterinary expertise, public health measures, and environmental factors all work together to prevent the spread of disease.

Throughout the book, case studies and real-world examples are used to illustrate key concepts and practices. These case studies not only showcase the practical application of theoretical knowledge but also highlight the diverse range of infectious diseases that affect different species, from domesticated animals to wild species in diverse ecosystems.

The final sections of the book focus on emerging infectious diseases and the future of veterinary medicine in combating infectious threats. It looks at the role of new technologies in diagnosis and treatment, as well as the increasing importance of global surveillance and collaboration in preventing pandemics and other widespread outbreaks.

Veterinary Infectious Diseases: Identification and Treatment is an essential resource for anyone involved in the care of animals. Whether you are a veterinarian in clinical practice, a researcher in veterinary medicine, or a student embarking on a career in the field, this book provides the critical information and practical advice needed to tackle the challenges posed by infectious diseases in animals. With a comprehensive approach that combines scientific understanding with clinical application, this book will help ensure the continued health and well-being of animals and protect public health from the threats posed by infectious diseases.

Ruchi Sandilya

Utkarsh Rajput

CONTENTS

1

ANIMAL DISEASES: OVERVIEW

Infectious diseases are transmitted between hosts by a variety of mechanisms, including direct, airborne and vector-borne transmission. Control of animal-to-animal transmission of disease agents is a key concept in infectious disease epidemiology; however, a more sensible approach might be to prevent the types of contact that lead to transmission in the first place. In humans, it is often difficult to prevent contacts, particularly with the ease of long-distance travel .

Animal disease, an impairment of the normal state of an animal that interrupts or modifies its vital functions. Concern with diseases that afflict animals dates from the earliest human contacts with animals and is reflected in early views of religion and magic. Diseases of animals remain a concern principally because of the economic losses they cause and the possible transmission of the causative agents to humans. The branch of medicine called veterinary medicine deals with the study, prevention, and treatment of diseases not only in domesticated animals but also in wild animals and in animals used in scientific research.

The prevention, control, and eradication of diseases of economically important animals are agricultural concerns. Programs for the control of diseases communicable from animals to man, called zoonoses, especially those in pets and in wildlife, are closely related to human health. Further, the diseases of animals are of increasing importance, for a primary public-health problem throughout the world is animal-protein deficiency in the diet of humans. Indeed, both the United Nations Food and Agricultural Organization (FAO) and the World Health Organization (WHO) have been attempting to solve the problem of protein deficits in a world whose human population is rapidly expanding. However, in livestock and animals, movements can be the subject of legislation or strict controls and there is a real opportunity to reduce disease transmission. The importance of animal movements is, of course, well understood and international regulations exist to mitigate the risks involved . In spite of these regulations, outbreaks occur regularly as a result of both legal and illegal animal movements.

GENERAL CONSIDERATIONS

Historical background

Historical evidence, like that from currently developing nations, indicates that veterinary medicine originally developed in response to the needs of pastoral and agricultural man along with human medicine. It seems likely that a veterinary profession existed throughout a large area of Africa and Asia from at least 2000 bce. Ancient Egyptian literature includes monographs on both animal and human diseases.

Evidence of the parallel development of human and veterinary medicine is found in the writings of Hippocrates on medicine and of Aristotle, who described the symptomatology and therapy of the diseases of animals, including man. Early Greek scholars, noting the similarities of medical problems among the many animal species, taught both human and veterinary medicine. In the late 4th century bce, Alexander the Great designed programs involving the study of animals, and medical writings of the Romans show that some of the most important early observations on the natural history of disease were made by men who wrote chiefly about agriculture, particularly the aspect involving domesticated animals.

Most of the earliest suggestions of relationships between human health and animal diseases were part of folklore, magic, or religious practice. The Hindu's concern for the well-being of animals, for example, originated in his belief in reincarnation. From the pre-Christian Era to about 1500, the distinctions between the practices of human and veterinary medicine were not clear-cut; this was especially true in the fields of obstetrics and orthopedics, in which animal doctors in rural areas often delivered babies and set human-bone fractures. It was realized, however, that training in one field was inadequate for practicing in the other, and the two fields were separated.

Veterinary literature from the civilizations of Greece and Rome contains reference to "herd factors" in disease; contagion within groups of animals kept together, therefore, was recognized, and both quarantine and slaughter were used to control outbreaks of livestock diseases. Rinderpest (cattle plague) was the most important livestock disease from the 5th century until control methods were developed. Serious outbreaks of the disease prompted the founding of the first veterinary college (École Nationale Vétérinaire), in Lyon, France, in 1762. Many aspects of animal diseases are best understood in terms of population or herd phenomena; for example, herds of livestock, rather than individual animals, are vaccinated against specific diseases, and housing, nutrition, and breeding practices are related to the likelihood of illness in the herd.

He work of Pasteur was of fundamental significance to general medicine and to agriculture. Veterinarians became concerned with foods of animal origin after the discovery of microorganisms and their identification with diseases in man and other animals. Efforts were directed toward protecting humans from diseases of animal origin, primarily those transmitted through meat or dairy products. Modern principles of food hygiene, first established for the dairy and meat-packing industries in the 19th and early 20th centuries, have been generally applied to other food-related industries. The veterinary profession, especially in Europe, assumed a major role in early food-hygiene programs.

- Spread of disease in animals
- The main causes of disease
- Disease can be classified as acute or chronic.
- An acute disease starts quickly and lasts for a short period when the animal either recovers or dies.
- A chronic disease lasts for a long time and weakens the animal.
- Diseases are said to be infectious (will spread from one animal to another) or noninfectious (will not spread from one animal to another).
- Non-infectious diseases can be caused by poor feed and the lack of minerals, salts and vitamins that the body needs.
- Non-infectious disease can also be caused by poisoning with chemicals or plants, by cuts, burns and broken bones.
- Some diseases pass from the parent to the young (hereditary).

Many non-infectious diseases are chronic but they can be acute. They can cause large losses of meat, milk and wool. Working (draught) animals do not work well and the rate of reproduction can be low with the young being born dead or dying before they are weaned.

Chronic diseases are often thought to be “normal” but when the cause is known and eliminated production can be greatly increased.

Infectious diseases are caused when the body is attacked by tiny living germs.

Spread of disease

Infectious diseases can be spread by:

Direct contact between animals.

Germs in feed and water.

By faeces and urine from sick animals.

By flies, ticks, lice and fleas.

By dirty housing or shelters.

Young and old animals become infected more easily.

Preventing infectious diseases

Animals, like humans, must be clean in order to be healthy. The animal must be provided with clean feed, water, bedding and shelter.

Sick animals should be kept separate from the others.

Some diseases can be cured by drugs.

Vaccination can protect animals against some diseases.

Dead animals and waste should be disposed of.

Preventing non-infectious diseases

The chronic non-infectious disease may not be recognized as a disease.

The affected animals may not die but will not produce as much milk, meat or wool, or work as well as could be expected.

If we continually look for ways of improving feed, water, mineral and vitamin supplies we will find the way to control the non-infectious diseases.

This will lead to greater production of wool, meat and milk, draught animals will be stronger and more young will be produced.

Poultry will produce more eggs and get fatter.

Livestock disease forewarning

Livestock sector plays a critical role in the welfare of India's rural population. This enterprise provides a flow of essential food products, draught power, manure, employment, income, and export earnings. As it is an important component in poverty alleviation programmes, continuous emphasis is being laid on this sector for enhancing the quality of the primary and secondary products, which in turn demands safe animal health for better products. Therefore, livestock development programmes cannot succeed unless a well-organized animal health service is built up and protection of livestock against diseases and pests particularly against the deadly infections is assured.

India has achieved eradication of Rinderpest (RP), Contagious Bovine Pleuro Pneumonia (CBPP), AHS and Dourine. However, there are several other infectious and non-infectious diseases prevailing in the country causing huge economic loss annually. Prevention, control and eradication of the animal diseases need a thorough understanding of the epidemiology as well their economic impact.

Early warning of disease incidence or outbreaks and the capacity of prediction of risk of spread to new areas is an essential pre-requisite for the effective containment and control of epidemic animal diseases.

National Animal Disease Referral Expert System (NADRES)

National Animal Disease Referral Expert System (NADRES) is an interactive, dynamic, relational online animal disease forewarning system that has been developed with overall aim to improve the early warning and response capacity to animal disease threats in the country for the benefit of farmers and policy makers in Animal Husbandry department.

National Institute of Veterinary Epidemiology and Disease Informatics (NIVEDI) with inputs from 31 centers spread across the country has developed the NADRES.

ICAR - NIVEDI has identified 13 priority diseases, based on the past incidence patterns and has built a strong database of these diseases. The database, which is backbone of the NADRES, is used for providing monthly livestock disease forewarning, which is compiled as monthly bulletins to alert the animal husbandry departments, both at the National/state level, to take appropriate control measures.

FOREWARNING METHODOLOGY

Materials

Livestock disease data - Previous 10 years livestock disease outbreak data retrieved from the NADRES database linked with Risk factors data.

Livestock population data : District wise livestock population data from 19th Livestock census (2012)

Meteorological data : Variables such as precipitation (mm/month), pressure (millibar), relative humidity (%), sea level pressure (millibar), minimum temperature

(° C), maximum temperature (° C), wind speed(m/s), vapour pressure (hPa), soil moisture (%), perceptible water (mm), potential evaporation transpiration (mm/ day) and cloud (%) are extracted from NCEP - National centre Environmental Prediction / IMD - Indian meteorological Database / NICRA - National Innovation Climate Resilient Agriculture and other sources for the past five years. Monthly average for the past five years is calculated and used.

Remote sensing data : Remote sensing variables such as NDVI - Normalised difference vegetation index, EVI - Enhanced vegetation index and LST - Land surface temperature are calculated using MODIS LANDSAT/IRS satellite images for the past five years. Monthly average for the past five years is calculated and used.

Statistical model

Disease outbreak is predicted by Generalised Linear Model (Logistic Regression) from the master chart containing the weather parameters, remote sensing variables and livestock population or densities using a R programme and the probability of disease outbreak is categorised in 6 risk levels - No risk (NR), Very low risk (VLR), Low risk (LR), Moderate risk (MR), High risk (HR) and Very high risk (VHR) for enabling the stake holders to take appropriate control measures by suitably allocating available resources.

Animals and Mechanisms of Disease Transmission

A diverse range of microbial pathogens can be transmitted from domestic and wild animals to human populations. These include viruses, bacteria, parasites, fungi, and prions. Strictly defined, not all infectious diseases common to animals and humans are zoonotic, and they can contract pathogens from the same sources, i.e., soil, water, plants, and invertebrates. For instance, acquisition of common human pathogens [Escherichia coli, Enterococci spp.] from animal food source with multiple antimicrobial resistance profile due to the liberal usage of antibiotics in animal feed may not be considered as zoonoses. Zoonotic diseases are due to transmissible infectious agents that affect more than one animal species, including humans, and cause clinical or subclinical infections.

The resurgence of zoonotic infectious diseases in the past two decades globally is of major concern. These diseases have resulted in significant morbidity and mortality on a large segment of the global human population, with over a billion afflicted resulting in millions of annual deaths . Moreover, outbreaks of zoonoses have had adverse impact on regional economies as a result of huge financial burden on the affected communities, but also indirectly affecting commerce or trade with more affluent consumer countries from the fear of spreading the affliction to their population. A systematic survey in 2007 estimated that in 1399 species of human pathogens, 87 of which were first reported in humans since 1980, most were viruses of animal reservoirs .

Socioecology Factors

Human-related factors include the results of industrialization and expansion of communities to accommodate the global population explosion. Developmental progress with clearing of forests for new roads, residences, towns, and farmlands impinge on wildlife ecology. Moreover, human intrusion on animal ecosystem is influenced by globalization of trade, alteration of farming and food chain practices, increased hunting and pet ownership, ecotourism, and expansion of culinary practice . Human exposures to animals and wildlife may be affected by changes in political regimes , conflict and wars, famine, mass migration and loosening of border controls, and breakdown of public health infrastructure. Furthermore, starvation and nutritional deficiencies usually result in a highly susceptible population to various diseases, including infections transmitted directly or indirectly from animals.

Pathogen-related factors are influenced by changes in the ecosystem and biodiversity that affect local fauna composition and quantity that may result in greater numbers of vectors and disease reservoirs/hosts, selection pressures for development of greater microbial resistance and virulence, and genetic variability. Climate and environmental factors are of increased concern in recent years, as global climate change with abnormal north-south hemisphere alteration of climate pattern can influence host-vector life cycles and change in fauna and ecology of animals and vectors.

Mechanisms of Transmission

There are several ways by which animals can transmit infectious agents to humans, and these include: direct contact with live animals or carcasses, indirect contact through animal products such as milk or eggs, intermediary transmission through vectors [fleas, mites, ticks, mosquitoes, etc.], and remote contact from exposure to contaminated waters, soil, and air. Direct contact with animals or carcasses can result in transmission of disease by a few different ways, most commonly by oral ingestion as in foodborne zoonoses, or accidentally from handling pets without proper hand sanitization.

Transmission of disease by animal bites, scratches, or mucosal exposure to infected secretions occurs relatively infrequently, such as in rabies and animal-related wound infections. Exposure to animal pathogens can be through inhalation of droplets of infected secretions, as believed to have occurred in China with local outbreaks of avian influenza arising from exposures to infected fowls or ducks in open markets. On rare occasions inhalation of microbial spores or contamination of the abraded or broken skin from exposure to contaminated animal hide can result in infection such as anthrax.

Exposures to animal excretions are common causes of unsuspected animal infectious pathogens and in sporadic cases the origin is usually undetected. Only in local outbreaks of diseases can the source of infection be detected by epidemiological investigations, as sometimes found from ingestion of contaminated vegetables from infected manure used for fertilization, as in listeria infection after ingestion of contaminated cold-slaw. Inhalation of aerosolized excreta from infected animals is another means of transmission, i.e., in " cave disease" that occurs in spelunkers after exploration of bat-infested caverns contaminated with spores or sporadic cases of hantavirus pulmonary disease from unwitting inhalation of rat excreta aerosolized with dust. Infected animal excretions may also transmit diseases through contamination of water used for drinking or bathing or from accidental skin exposure to puddles of water containing rat urine to produce leptospirosis.

Intermediary transmission by vectors represents a major mode of transmitting animal pathogens to humans. In most instances humans are incidental hosts and are not essential for maintaining the life cycle of the parasite or pathogen. Vector-transmitted zoonoses have been a scourge to humanity since antiquity and continue to plague human populations at present and for the foreseeable future. Chagas disease, transmitted by the " kissing bug" [triatomine], afflicted humans as early as 9000 years ago. The causative [discovered in 1909] can be found in at least 150 species of domestic and wild animals; and paleoparasitologists detected T. cruzi DNA from human mummies of ancient times . Vector-borne zoonoses continue to emerge in the modern era, for example, Lyme disease, and expand with greater resurgence as exemplified by the global expansion of dengue fever and West Nile virus disease .

ANIMAL DISEASE VIA THE FOOD CHAIN

Accurate estimates of the global burden of foodborne infectious diseases are difficult to attain and are grossly underestimated in countries with active surveillance. Even in developed affluent countries such as the United States [US], most foodborne diseases are unreported, and only a fraction of these cases have a microbiological etiology confirmation. Estimates of foodborne infections are even more difficult to derive in developing countries, where there are inadequate facilities and healthcare infrastructure and the global burden of foodborne zoonoses is the greatest. It is estimated that each year at least one-third of world's population is afflicted with foodborne infections and a large fraction from animal pathogens. International collaboration spearheaded by the World Health Organization [WHO] launched surveillance programs since 2000 to determine the burden of foodborne diseases in resource-poor countries and globally .

Previous estimates from the Center for Disease Control and Prevention [CDC] in 2011 determined that 31 major pathogens acquired in the US caused at least 9.4 million episode of foodborne illness each year, but could be >48 million cases . Common animal-derived pathogens such as nontyphoidal Salmonella spp., Campylobacter spp., and Toxoplasma gondii were the leading causes of hospitalization. It is surmised that about 30% of all emerging infectious diseases in the past 60 years were caused by pathogens commonly transmitted through food and were of animal origin .

There are more than 250 different foodborne diseases, the majority due to transmissible microbes [bacteria, viruses, parasites, and prions] and the remainder by chemicals and toxins . Although foodborne zoonoses can be asymptomatic and the majority present with acute self-limited disease, occasionally severe disease and fatality occur and chronic disability can be present. Infection may result from eating raw or undercooked animal products, or contaminated vegetables and fruits, or drinking contaminated water or milk . Moreover, infections and outbreaks can occur from prepared food contaminated by an ill person or asymptomatic carrier. It is often difficult to trace animal-mediated infectious diseases from non-meat products as shown by recent experiences. This is exemplified by the outbreak in Europe of a novel Shiga toxin-producing E. coli 0104:H4 from contaminated sprouts originating in Germany, resulting in 50 deaths, and the outbreak in the US with Listeria-contaminated cantaloupes which killed 29 people.

Survey of animal diseases

Infectious and noninfectious diseases

Diseases may be either infectious or noninfectious. The term infection, as observed earlier, implies an interaction between two living organisms, called the host and the parasite. Infection is a type of parasitism, which may be defined as the state of existence of one organism (the parasite) at the expense of another (the host). Agents (e.g., certain viruses, bacteria, fungi, protozoans, worms, and arthropods) capable of producing disease are pathogens. The term pathogenicity refers to the ability of a parasite to enter a host and produce disease; the degree of pathogenicity—that is, the ability of an organism to cause infection—is known as virulence. The capacity of a virulent organism to cause infection is influenced both by the characteristics of the organism and by the ability of the host to repel the invasion and to prevent injury. A pathogen may be virulent for one host but not for another. Pneumococcal bacteria, for example, have a low virulence for mice and are not found in them in nature; if introduced experimentally into a mouse, however, the bacteria overwhelm its body defenses and cause death.

Many pathogens (e.g., the bacterium that causes anthrax) are able to live outside the animal's body until conditions occur that are favourable for entering and infecting it. Pathogens enter the body in various ways—by penetrating the skin or an eye, by being eaten with food, or by being breathed into the lungs. After their entry into a host, pathogens actively multiply and produce disease by interfering with the functions of specific organs or tissues of the host.

Before a disease becomes established in a host, the barrier known as immunity must be overcome. Defense against infection is provided by a number of chemical and mechanical barriers, such as the skin, mucous membranes and secretions, and components of the blood and other body fluids. Antibodies, which are proteins formed in response to a specific substance (called an antigen) recognized by the body as foreign, are another important factor in preventing infection. Immunity among animals varies with species, general health, heredity, environment, and previous contact with a specific pathogen.

As certain bacterial species multiply, they may produce and liberate poisons, called exotoxins, into the tissues; other bacterial pathogens contain toxins, called endotoxins, which produce disease only when liberated at the time of death of the bacterial cell. Some bacteria, such as certain species of Clostridium and Bacillus, have inactive forms called spores, which may remain viable (i.e., capable of developing into active organisms) for many years; spores are highly resistant to environmental conditions such as heat, cold, and chemical compounds called disinfectants, which are able to kill many active bacteria.

The term infestation indicates that animals, including spiny-headed worms (Acanthocephala), roundworms (Nematoda), flatworms (Platyhelminthes), and arthropods such as lice, fleas, mites, and ticks, are present in or on the body of a host. An infestation is not necessarily parasitic.

Noninfectious diseases are not caused by virulent pathogens and are not communicable from one animal to another. They may be caused by hereditary factors or by the environment in which an animal lives. Many metabolic diseases are caused by an unsuitable alteration, sometimes brought about by man, in an animal's genetic constitution or in its environment. Metabolic diseases usually result from a disturbance in the normal balance of the physiological mechanisms that maintain stability, or homeostasis. Examples of metabolic diseases include overproduction or underproduction of hormones, which control specific body processes; nutritional deficiencies; poisoning from such agents as insecticides, fungicides, herbicides, fluorine, and poisonous plants; and inherited deficiencies in the ability to synthesize active forms of specific enzymes, which are the proteins that control the rates of chemical reactions in the body.

Excessive inbreeding (i.e., the mating of related animals) among all domesticated animal species has resulted in an increase in the number of metabolic diseases and an increase in the susceptibility of certain animals to infectious diseases.

Zoonoses

As stated previously, zoonoses are human diseases acquired from or transmitted to any other vertebrate animal. Zoonotic diseases are common in currently developing countries throughout the world and constitute, with starvation, the major threat to human health. More than 150 such diseases are known.

Zoonoses may be separated into four principal types, depending on the mechanisms of transmission and epidemiology. One type includes the direct zoonoses, such as rabies and brucellosis, which are maintained in nature by one vertebrate species. The transmission cycle of the cyclozoonoses, of which tapeworm infections are an example, requires at least two different vertebrate species. Both vertebrate and invertebrate animals are required as intermediate hosts in the transmission to humans of metazoonoses; arboviral and trypanosomal diseases are good examples of metazoonoses. The cycles of saprozoonoses (for example, histoplasmosis) may require, in addition to vertebrate hosts, specific environmental locations or reservoirs.

Most animals that serve as reservoirs for zoonoses are domesticated and wild animals with which man commonly associates. People in occupations such as veterinary medicine and public health, therefore, have a greater exposure to zoonoses than do those in occupations less closely concerned with animals.

In addition to the numerous human diseases spread by contact with the parasitic worm helminth and by contact with arthropods, many diseases are transmitted by the bites and venom of certain animals; poisonous or diseased food animals also transmit diseases. Dog bites may seriously injure tissues and also can transmit bacterial infections and rabies, a disease of viral origin. The bite of a diseased rat may transmit any of several diseases to man, including plague, salmonellosis, leptospirosis, and rat-bite fevers. Cat scratch disease may be transmitted through cat bites, and the deadly herpes B virus can spread by monkey bites. The bites of venomous snakes and fish account for considerable human discomfort and death. About 200 of the 2,500 known species of snakes can cause human disease. One estimate for snakebite deaths worldwide is 30,000 to 40,000 per year, the vast majority of them in Asia. Poisonous wild animals inadvertently used for food include animals harbouring the anthrax bacillus and those containing the causative agents of salmonellosis, trichinosis, and fish-tapeworm infection. The flesh of various types of fish is toxic to man. Japanese puffers, for example, contain the poisonous chemical compound tetrodotoxin; scombroid fish harbour Proteus morganii, which causes gastrointestinal diseases; and mullet and surmullet can cause nervous disturbances.

Approaches to the control of zoonoses differ according to the type under consideration. Because the majority of direct zoonoses and cyclozoonoses and some saprozoonoses are most effectively controlled by techniques involving the animal host, methods used to combat these diseases are almost entirely the responsibility of veterinary medicine. A good example is the elimination of stray dogs, for they are an important factor in the control of zoonoses such as rabies, hydatid disease, and visceral larva migrans.

In addition, the control of diseases such as brucellosis and tuberculosis in cattle involves a combination of methods—mass immunization, diagnosis, slaughter of infected animals, environmental disinfection, and quarantine. Several supportive measures for the control of disease are useful in some cases. Air-sanitation measures are helpful in direct zoonoses in which human illness is spread by droplets or dust, and zoonotic infections that are spread through a fluid medium, such as water or milk, sometimes can be controlled. Heat, cold, and irradiation are effective in killing the immature forms of Trichinella spiralis, the causative agent of trichinosis, in meat; and certain antibiotic drugs help to prevent deterioration of food.

The control of metazoonoses may be directed at the infected vertebrate hosts, at the infected invertebrate host, or at both. Particularly effective in this instance has been the use of chemical insecticides to attack the invertebrate carriers of specific infections, even though several difficulties have been encountered—for example, the inaccessibility of the invertebrate to the chemicals, which occurs with organisms that breed in swiftly flowing waters or in dense vegetation, and the development of insecticide resistance by the organisms. Insecticides are used to destroy the mosquitoes that spread malaria (Anopheles). Mechanical filters placed across irrigation ditches help to prevent the dissemination of the snails that transmit Schistosoma mansoni, a parasitic flatworm.

DISEASE PREVENTION, CONTROL, AND ERADICATION

Prevention is the first line of defense against disease. At least four preventive techniques are available for use in the prevention of disease in an animal population. One is the exclusion of causative agents of disease from specific geographic areas, or quarantine. A second preventive tool utilizes control methods such as immunization, environmental control, and chemical agents to protect specific animal populations from endemic diseases, diseases normally present in an area. The third preventive measure concerns the mass education of people about disease prevention. Finally, early diagnosis of illness among members of an animal population is important so that disease manifestations do not become too severe and so that affected animals can be more easily managed and treated.

Quarantine—the restriction of movement of animals suffering from or exposed to infections such as bluetongue and scrapie (in sheep), foot-and-mouth disease (in cattle), and rabies (in dogs)—is one of the oldest tools of preventive medicine. It was applied to domesticated animals as early as Roman times. The establishment of international livestock quarantine in the United States in 1890 provided for the holding of all imported cattle, sheep, and swine at the port of entry for 90, 15, and 15 days, respectively. In this way, such diseases as Nairobi sheep disease, surra, and infections caused by Brucella melitensis were eliminated or excluded from the United States, but international quarantine barriers did not prevent the entry of bluetongue, scrapie, and the tick Rhipicephalus evertsi, which is a carrier for several animal diseases. On the other hand, long-term quarantine of all dogs entering Great Britain has been effective since its initiation in 1919 (the quarantine also includes cats). It is possible that aircraft may pose new problems regarding livestock-disease quarantine since many disease carriers (e.g., insects and viruses) may be accidentally brought by plane into a country.

Mass immunization as a preventive technique has the advantage of allowing the resistant animal freedom of movement, unlike environmental control, in which the animal is confined to the controlled area; immunization may, however, provide only short-lived and partial protection. Mass-inoculation techniques against diseases such as Newcastle disease in chickens and distemper in mink and dogs have been successful. Animal diseases have been prevented by methods involving environmental control, including the maintenance of safe water supplies, the hygienic disposal of animal excrement, air sanitation, pest control, and the improvement of animal housing.

One specific environmental program, called the portable-calf-pen system, involves routine movement of the pens to avoid a concentration of specific pathogens in them. Other programs involve the utilization of automatic and sanitary watering and feeding equipment and buildings with environmental controls. The use of chemical compounds to prevent illness (chemoprophylaxis) includes a variety of pesticides, which are used to kill insects that transmit diseases, and substances either used internally or applied to the animal's body to prevent the transmission or the development of a disease. An example is the use of sulfonamide drugs in the drinking water of poultry to prevent coccidiosis. Environmental-control methods in the poultry industry have resulted in the most efficient means of poultry production developed thus far.

The early detection of a disease in a population of animals—a herd of cattle, for example—is particularly useful in controlling certain chronic infectious diseases, such as mastitis, brucellosis, and tuberculosis, as well as certain noninfectious diseases such as bloat. Laboratory tests—the agglutination test in pullorum disease, the tuberculin skin test for tuberculosis, the examination of feces for eggs of specific parasites, the physical and chemical tests performed on milk to diagnose bovine mastitis—are used for the early detection of diseases in an animal population.

Methods of disease control and eradication have been successful in various countries. In the United States, for example, the test-and-slaughter technique, in which simple tests are used to confirm the existence of diseased animals that are then slaughtered, has been of great value in controlling infectious and hereditary diseases, including dourine, a venereal disease in horses, fowl plague, and foot-and-mouth disease in cattle and deer. Bovine tuberculosis has been eliminated from Denmark, Finland, and The Netherlands and reduced to a low level in various other countries, including Great Britain, Japan, the United States, and Canada, by the test-and-slaughter method. Many infectious diseases have been eradicated from Great Britain—sheep pox, rinderpest, pleuropneumonia, glanders, and rabies.

Diseases eliminated from Australia by a combination of methods—control of agents that carry disease, the test-and-slaughter technique, the use of chemical agents, and, more recently, biological control—include hog cholera, rinderpest, scrapie, glanders, surra, rabies, and foot-and-mouth disease.

In biological control, enemies of the agents that transmit the disease, enemies of the reservoir host, or a specific parasite are introduced into the environment. If a natural enemy of the tsetse fly could be found, for example, African sleeping sickness in man and trypanosomiasis in cattle could be controlled in West Africa. Successful biological control of the European-rabbit population in Australia has been accomplished through the use of the myxomatosis virus, which is transmitted by mosquitoes and causes the formation of malignant tumours. Although the Brazilian white rabbit is relatively unaffected by the virus, it causes rapid death in the European rabbit.

The elimination of the European rabbit in France by the virus was accompanied by a decrease in tick-borne typhus in people, suggesting that the rabbit may be a significant intermediate host for the causative agent, Rickettsia conorii. Screwworms, an immature form of the fly Cochliomyia hominivorax, have been eradicated in the United States by the release of more than 3,000,000,000 sterilized males.

Disease control and elimination programs require many sophisticated techniques in addition to diagnosis and the slaughter of affected animals. They include: the control of insects known to transmit diseases; the cooperation of animal owners; the development through research of new diagnostic tests for use on large populations; the eradication of animal species from areas in which they are known to transmit disease; sterilization of strains of animals known to carry inheritable metabolic diseases; and effective meat inspection.

2

COMMON ANIMAL DISEASES AND THEIR TREATMENT

Animal viruses, unlike the viruses of plants and bacteria, do not have to penetrate a cell wall to gain access to the host cell. Non-enveloped or "naked" animal viruses may enter cells in two different ways. When a protein in the viral capsid binds to its receptor on the host cell, the virus may be taken inside the cell via a vesicle during the normal cell process of receptor-mediated endocytosis. An alternative method of cell penetration used by non-enveloped viruses is for capsid proteins to undergo shape changes after binding to the receptor, creating channels in the host cell membrane.

The viral genome is then "injected" into the host cell through these channels in a manner analogous to that used by many bacteriophages. Enveloped viruses also have two ways of entering cells after binding to their receptors: receptor-mediated endocytosis and fusion. Many enveloped viruses enter the cell by receptor-mediated endocytosis in a fashion similar to some non-enveloped viruses. On the other hand, fusion only occurs with enveloped virions. These viruses, which include HIV among others, use special fusion proteins in their envelopes to cause the envelope to fuse with the plasma membrane of the cell, thus releasing the genome and capsid of the virus into the cell cytoplasm.

After making their proteins and copying their genomes, animal viruses complete the assembly of new virions and exit the cell. Using the example of HIV, enveloped animal viruses may bud from the cell membrane as they assemble themselves, taking a piece of the cell's plasma membrane in the process. On the other hand, non-enveloped viral progeny, such as rhinoviruses, accumulate in infected cells until there is a signal for lysis or apoptosis, and all virions are released together.

Animal viruses are associated with a variety of human diseases. Some of them follow the classic pattern of acute disease, where symptoms worsen for a short period followed by the elimination of the virus from the body by the immune system with eventual recovery from the infection. Examples of acute viral diseases are the common cold and influenza. Other viruses cause long-term chronic infections, such as the virus causing hepatitis C, whereas others, like herpes simplex virus, cause only intermittent symptoms. Still other viruses, such as human herpes viruses 6 and 7, which in some cases can cause the minor childhood disease roseola, often successfully cause productive infections without causing any symptoms at all in the host; these patients have an asymptomatic infection.

In hepatitis C infections, the virus grows and reproduces in liver cells, causing low levels of liver damage. The damage is so low that infected individuals are often unaware that they are infected, with many infections only detected by routine blood work on patients with risk factors such as intravenous drug use. Since many of the symptoms of viral diseases are caused by immune responses, a lack of symptoms is an indication of a weak immune response to the virus. This allows the virus to escape elimination by the immune system and persist in individuals for years, while continuing to produce low levels of progeny virions in what is known as a chronic viral disease. Chronic infection of the liver by this virus leads to a much greater chance of developing liver cancer, sometimes as much as 30 years after the initial infection.

As mentioned, herpes simplex virus can remain in a state of latency in nervous tissue for months, even years. As the virus "hides" in the tissue and makes few if any viral proteins, there is nothing for the immune response to act against; immunity to the virus slowly declines. Under certain conditions, including various types of physical and psychological stress, the latent herpes simplex virus may be reactivated and undergo a lytic replication cycle in the skin, causing the lesions associated with the disease. Once virions are produced in the skin and viral proteins are synthesized, the immune response is again stimulated and resolves the skin lesions in a few days by destroying viruses in the skin. As a result of this type of replicative cycle, appearances of cold sores and genital herpes outbreaks only occur intermittently, even though the viruses remain in the nervous tissue for life. Latent infections are common with other herpes viruses as well, including the varicella-zoster virus that causes chickenpox. After having a chickenpox infection in childhood, the varicella-zoster virus can remain latent for many years and reactivate in adults to cause the painful condition known as "shingles".

ANTHRAX

Anhrax, a highly infectious and fatal disease of cattle, is caused by a relatively large spore-forming rectangular shaped bacterium called Bacillus anthracis. Anthrax causes acute mortality in ruminants. The bacteria produce extremely potent toxins which are responsible for the ill effects, causing a high mortality rate. Signs of the illness usually appear 3 to 7 days after the spores are swallowed or inhaled. Once signs begin in animals, they usually die within two days.

Hoofed animals, such as deer, cattle, goats, and sheep, are the main animals affected by this disease. They usually get the disease by swallowing anthrax spores while grazing on pasture contaminated (made impure) with anthrax spores. Inhaling (breathing in) the spores, which are odorless, colorless, and tasteless, may also cause infection in animals and people.

Symptoms:

Sudden death (often within 2 or 3 hours of being apparently normal) is by far the most common sign;

Very occasionally some animals may show trembling, a high temperature

Difficulty breathing, collapse and convulsions before death. This usually occurs over a period of 24 hours;

After death blood, may not clot, resulting in a small amount of bloody discharge from the nose, mouth and other openings

Treatment and control

Due to the acute nature of the disease resulting in sudden death, treatment is usually not possible in animals even though Anthrax bacilli are clines. Treatment is of use in cases showing sub-acute form of the disease.

In most cases, early treatment can cure anthrax. The cutaneous (skin) form of anthrax can be treated with common antibiotics.

Preventive measures:

Regular annual vaccination of animals in endemic areas will prevent the disease from occurring.

Vaccination may be carried out at least a month prior to expected disease occurrence in endemic areas.

Never open a carcass of an animal suspected to have died from anthrax.

Contact a veterinarian immediately if the following symptoms are seen and seek advice on control measures to be adopted.

Fever (106-108°F), loss of appetite, depression and dullness

Suspended rumination3. Rapid pulse and heart rates

Difficult breathing (dyspnoea)

Lameness in affected leg

Crepitation swelling over hip, back & shoulder

Swelling is hot & painful in early stages whereas cold and painless inter.

Recumbency (prostration) followed by death within 12-48 hrs.

BLACK QUARTER (BLACK-LEG)

It is an acute infectious and highly fatal, bacterial disease of cattle. Buffaloes, sheep and goats are also affected. Young cattle between 6-24 months of age, in good body condition are mostly affected. It is soil-borne infection which generally occurs during rainy season. In India, the disease is sporadic (1-2 animal) in nature.

Causal organism: it is a bacterial disease caused by Clostridium chauvoei

Symptoms

Fever (106-108°F), Loss of appetite, Depression and dullness

Suspended rumination

Rapid pulse and heart rates

Difficult breathing (dyspnoea)

Lameness in affected leg

Crepitation swelling over hip, back & shoulder

Swelling is hot & painful in early stages whereas cold and painless inter.

Recumbency (prostration) followed by death within 12-48 hrs.

Treatment:

Early treatment can be possible to complete cure of the animal.

Consult with veterinarian immediately.

Ethnovet practice

The following measure is to be taken up in the month of May / June every year. Exudates of thirugukalli (Euphorbia tirucalli), kodikalli (Sareostemma brevistigma), aththi (Ficus racemosa), banyan tree (Ficus bengalensis), madara (Calotropis gigantea) are taken at the rate of 1 to 15 drops each in a stainless-steel vessel and mixed with 50 ml of sesame oil and ragi flour are added and made into a paste. This paste is applied as dot (coin size) in each animal in the groin region. (the above material may be used for about 50 animals).

FOOT AND MOUTH DISEASE

The foot-and-mouth disease is a highly communicable disease affecting cloven-footed animals. It is characterized by fever, formation of vesicles and blisters in the mouth, udder, teats and on the skin between the toes and above the hoofs. Animals recovered from the disease present a characteristically rough coat and deformation of the hoof. In India, the disease is widespread and assumes a position of importance in livestock industry. The disease spreads by direct contact or indirectly through infected water, manure, hay and pastures. It is also conveyed by cattle attendants. It is known to spread through recovered animals, field rats, porcupines and birds.

Symptoms

- fever with 104-105o F
- profuse salivation - ropes of stringy saliva hangs from mouth
- vesicles appear in mouth and in the inter digital space
- lameness observed
- cross bred cattle are highly susceptible to it

Treatment

the external application of antiseptics contributes to the healing of the ulcers and wards off attacks by flies.

a common and inexpensive dressing for the lesions in the feet is a mixture of coal-tar and copper sulphate in the proportion of 5:1.

Precautions

heavy milch animals and exotic breeds of cattle bred for milk should be protected regularly.

It is advisable to carry out two vaccinations at an interval of six months followed by an annual vaccination programme.

Isolation and segregation of sick animals. It should be informed immediately to the veterinary doctor disinfection of animal sheds with bleaching powder or phenol attendants and equipment's for sick animals should be ideally separate the equipment's should be thoroughly sanitized proper disposal of left over feed by the animal proper disposal of carcasses control of flies.

Ethnovet prevention practice:

When there is a outbreak in the nearby villages /surroundings take tulasi (Ocimum sp) leaves 100 gm, a pinch of common salt and turmeric rhizome 2 pieces and grind them. This has to be squeezed to obtain extract and administered orally. The residues left over can be used for smearing over the mouth region, foot region. This is repeated.

RABIES (MAD DOG DISEASE)

Rabies is a disease of dogs, foxes, wolves, hyaenas and in some places, it is a disease of bats which feed on blood.

The disease is passed to other animals or to people if they are bitten by an animal with rabies. The germs which cause rabies live in the saliva of the sick (rabid) animal. This is a killer disease but not every dog which bites is infected with rabies.

When the rabid animal bites another animal or human, the germs which live in its saliva pass into the body through the wound caused by the bite. The germs travel along the nerves to the brain. The time between the bite and the first appearance of signs that the bitten animal or human has been infected can take from 2 to 10 weeks or more. The time taken depends on the distance of the bite from the brain. If the bite is on the face or head, the bitten animal or human will quickly show signs, but if the bite is on the leg it will take much longer for signs to develop.

General signs of rabies

You should first look for the marks of the bite and discover where and when the animal was bitten. All rabid animals show similar signs in the beginning.

They change their normal behaviour and behave very strangely.

They stop eating or drinking.

Male animal will try to mate (mount) other animals.

There is no change in the body temperature.

These signs will continue for 3 to 5 days. Then, before it dies, the animal will develop one or the other of two types of the disease:

The furious (mad) type of the disease makes the animal aggressive and it will bite anything.

The quiet (dumb) type when the animal is quiet and does not move.

Rabies in the dog

Dogs show either of the two types of rabies.

a dog with the dumb or quiet type of the disease cannot move. It looks as if it has a bone stuck in the mouth and saliva drips from the mouth.

rabies in the dog lasts about 10 days before the animal dies. If the animal does not die after this length of time then it may not be suffering from rabies.

Rabies in Sheep, Goats and Cattle

Rabies is characterised by the animals becoming restless and excited. They may bite themselves and saliva drips from the mouth. The most important sign in cattle is that the animal bellows (calls) very frequently and with strange sound. The animals will become paralysed and die.

Rabies in the horse and camel

The horse will show the furious (mad) type of the disease. It will kick and bite and show signs similar to colic. The animal will die after paralysis of the back legs.

In the camel the signs of rabies are similar to those shown by an animal in the rut.

What to do with a biting dog

Remember that not every dog which bites has rabies. If the dog belongs to somebody ask the owner about its normal behaviour. If the dog is showing signs of rabies you must inform your veterinary officer immediately. The dog must be shot and if it has bitten anybody, they must be taken to a hospital immediately for vaccination.

Control of rabies

Dogs in your community can be vaccinated against rabies. You should ask your veterinary service about vaccination against rabies. If there is an outbreak of rabies, the livestock in your community can be vaccinated too.

Treatment (ethnovet practices) :

Leaves of chirchra (Achyranthes aspera) 100gm and onion 50 gm are ground well and smeared over the bitten place. The extract of these ingredients is administered orally twice in a day.

Blue tongue

Bluetongue, a disease which is transmitted by midges, infects domestic and wild ruminants and also camelids, however sheep are particularly badly affected. Cattle, although infected more frequently than sheep, do not always show signs of disease. Virus spreads between animals occurs via the midges of Cullicoides species.

The likelihood of mechanical transmission between herds and flocks, or indeed within a herd or flock, by unhygienic practices (the use of contaminated surgical equipment or hypodermic needles) may be a possibility.

Clinical signs include

Sheep : eye and nasal discharges, drooling, high body temperature, swelling in mouth, head and neck, lameness and wasting of muscles in hind legs, haemorrages into or under skin, inflammation of the coronary band, respiratory problems, fever, lethargy.

In cattle : nasal discharge, swelling of head and neck, conjunctivitis, swelling inside and ulceration of the mouth, swollen teats, tiredness, saliva drooling, fever.

Note: a blue tongue is rarely a clinical sign of infection

Control

Inspect stock closely, particularly focusing on the lining of the mouth and nose and the coronary band (where the hoof stops and the skin starts). If an animal is suspected as having bluetongue, it must be reported as quickly as possible. Telephone your local animal health office immediately.

Preventive measures and treatment (ethovet)

Since the animal is not taking any feed the starvation may lead to death. So the animal has to be administered orally the following food. Banana fruits (one) smeared with sesame oil (50 ml) for 2 to 3 times. By this animal will recover little. However, this will not control the disease fully. Next the leaf pulp of "sothukathalai"(Aloe vera) has to be administered daily. Administering of Aloe vera has to be continued for more days till the animal fully recovers from this disease. By this treatment the infected animal will recover from the disease. The disease will not spread to other animals if all animals are administered with Aloe vera as a preventive treatment. Administering aloe vera also increases the body weight of animals as it is against all intestinal parasite.

POX

Epidemiology : sheep-pox is a highly contagious disease. It causes a mortality of 20 to 50 per cent in animals below the age of 6 months, and causes damage to the wool and skin in adults. Of the pock diseases, sheep-pox ranks only second to human small-pox in virulence. The disease is transmissible to in-contact goats but not to other species of animals. It, however, spreads slowly.

Symptoms : The disease in characterized by high fever, and symptoms of pneumonia and acute enteritis. Skin lesions appear particularly in parts free from wool, notably around the eyes, inner side of the thigh, udder and under surface of the tail. The internal organs such as trachea, lungs, kidneys and intestines are also affected. The disease results in emaciation and, as already mentioned, frequent deaths of affected animals.

Treatment, prevention and control

The diseased animal should be treated with palliatives. In the young ones nursing is more important than medication. The infected litter should be burnt and the bedding changed every day. Affected animals should be kept on soft diet. The ulcers on the skin should be washed with potassium permanganate lotion and dusted with boric acid; strict hygienic measures should be adopted.

Preventive measures and treatment (ethnovet)

External application of paste prepared by grinding neem leaves, tulsi leaves each 100 gm and turmeric powder- 50gm sprinkled with sufficient water. Continue for 3 to 5 days. Administer orally the same mixture by diluting with water.

Brucellosis of sheep

Transmission : The mode of entry is by ingestion or via conjunctiva. The aborted foetus, vaginal discharge and milk from infected goats contain a large number or organisms.

Symptoms in infected goats and sheep state of abortion may occur followed by a quiescent period during which a few abortions occur. The aborted animals do not breed. After 2 years or more another abortion storm is likely to occur.

Diagnosis, treatment and control

It is not possible to diagnose brucellosis on the basis of symptoms alone. The suspicion is aroused when humans in contact suffer from undulant fever and there is poor breeding record in goat herd and evidence of mastitis. The diagnosis can be done by the isolation of organisms and by serological tests.

There is no adequate treatment

This is based on hygiene, vaccination, testing and disposal. Good management practice is essential. Separate quarters should be provided for kidding. Immunization can be done with attenuated as well as killed vaccines. The test and disposal procedure is highly desirable.

TETANUS

This is an infectious, non-febrile disease of animals and man, and is characterized by spasmodic tetany and hyperaesthesia. This disease is prevalent all over the world.

Transmission : Infection takes place by contamination of wounds. Deep punctured wounds provide favourable conditions for the spores to germinate, multiply and produce toxin which is subsequently absorbed in the animal body. The micro-organism is present in soil and in animal faeces, and is carried into the wound by a penetrating object. The organism is present in the intestine of normal animals, and under some undetermined conditions multiplies rapidly and produces toxin in sufficient quantities to be absorbed and cause the disease.

Symptoms : The incubation period is generally 1-2 weeks but it may be as short as 3 days. Tetanus affects many species of domesticated animals but occurs particularly in horses and lambs; less frequently in adult sheep, goats, cattle, pigs, dog and cats; and rarely in poultry. The initial symptoms are mild stiffness and an unwillingness to move all the animals. More severe symptoms develop after 12-24 hours which are stiffness of limbs, neck, head, tail and twitching of muscles. The spasms develop in response to noise. In terminal stages ears are erect, nostrils dilated, nictitating membrane protruded. Mastication becomes very difficult because mouth cannot be opened, hence the name lockjaw.

Treatment : The treatment is carried out by first injecting antitoxin then treating the wound. Penicillin parenterally is beneficial. Muscular relaxation is achieved by injection of relaxants. The animal should be kept in a dark room and fed with the help of stomach tube.

Control : Proper hygiene and cleanliness at castration and other surgical procedures should be observed. Sheep should be given 2 injections based 3 weeks apart to develop a solid immunity.

LISTERIOSIS

Transmission : The organisms are excreted in the faeces, urine, aborted foetuses, uterine discharge and milk of infected animals. The organisms are sufficiently resistant to remain viable in animal and human faeces, sewage, soil, silage and dust foe several weeks and months. The blood sucking arthropods may spread infection since organisms have been isolated from cattle ticks and tabanid flies. Under natural conditions certain predisposing factors are related to clinical infection.

Symptoms : In farm animals the disease occurs towards the end of winter or early spring. The first signs of meningo- encephalitis are stiffness of neck, incoordinated movement of limbs and tendency to move in circles or to lean against a fence or wall. There may be paralysis of muscles of jaw and pharynx. Incoordination becomes progressively more severe until the animal can no longer stand.

The cattle which are not severely affected may survive. Abortions in cattle usually occur after 4-8 months of pregnancy and at a comparatively later stage in sheep. In pigs and horses, clinical signs are not common but may develop as encephalitis and septicaemia. In poultry, the disease usually causes sudden death, occasionally there are signs of torticollis, weakness and inco-ordination of the legs.

Treatment : Tetracyclines are very effective in meningo-encephalities of cattle less so in sheep. The recovery rate depends on the speed with which the treatment is commenced.

Control : When outbreaks occur all affected animals should be slaughtered and buried along with litter and bedding. The vaccines, living or killed, have little effect on the pathogenesis of infection under natural conditions, tetracycline's are very effective for treatment of listeriosis.

CAMPYLOBACTOR ABORTION (VIBRIOSIS)

Transmission : transmission occurs by coitus. The affected bulls carry the organisms in preputial cavity indefinitely. Mature cows and heifers also carry the infection for long periods. Infected semen from an infected bull is the important means of the disease. The organism survives low temperature used in semen storage.

Symptoms : infertility may cause become apparent only when the percentage of pregnancies in a dairy herd is low. The infertility rate in heifers is more than in cows. Abortions usually occur between fifth and sixth month of pregnancy. Infected bulls show no symptoms and their semen is normal. Healthy bulls become infected during coitus with diseased cow. Among sheep the disease is characterized by abortion occurring towards the end of gestation. Usually abortion is preceded by vaginal discharge for several days. The aborted foetus is edematous with petechial hemorrhages on serous surfaces and necrotic foci in the liver.

Control ; abortion rate can be reduced by antibiotic therapy, and particularly by using chlortetracycline and concurrently with the development of specific immunity. The use of killed vaccines may reduce the incidence of disease in a herd but does not eradicate the infection. The bulls can be treated by injecting antibiotic cream in the prepuce. There is no direct treatment of females.

JOHNE`S DISEASE

Johne`s disease is a specific chronic contagious enteritis of cattle, sheep, goat, buffaloes and occasionally of pigs. The disease is characterized by progressive emaciation and in cattle and buffaloes by chronic diarrhea and thickening of the intestine. Transmission under natural conditions the disease spread by ingestion of feed and water contaminated by the faeces of infected animals. The infection occurs mostly in the early month of life. The incubation period extends from 12 months to several years. The animal aged 3 to 6 years mostly suffer from the disease. Affected animals may not show clinical symptoms continue to discharge organisms in faeces. The organisms persist in pastures for about 1 year. The organisms are susceptible to sunlight, drying and high ph of soil; continuous contact of urine with faeces reduces the life of bacteria. In cattle clinical signs appear mainly during 2-6 years of age. The infected animals which are apparently healthy, often show clinical signs after parturition.

Treatment the organisms is more resistant to chemotherapeutic agents invitro than mycotuberculosis. Because of this the practical utility of treatment in clinical cases is poor.

Control the affected animal should be segregated and their faeces properly disposed off. Alivc vaccine has been developed. It reduces the incidence of clinical disease. It consists of a non-pathogenic strain of jhone`s bacillus with an adjuvant. The calves soon after birth are inoculated with vaccine subcontaneously. The vaccinated animals become reactors of jhonin. Vaccination is generally done in heavily infected herds.

Bovine ephemeral fever

Bovine ephemeral fever is an insect-transmitted, noncontagious, viral disease of cattle and water buffalo that is seen in Africa, the middle east, Australia, and Asia.

Etiology and epidemiology

Bovine ephemeral fever virus (befv) is classified as a member of the genus Ephemerovirus in the family Rhabdoviridae (single-stranded, negative sense rna)

The prevalence, geographic range, and severity of the disease vary from year to year, and epidemics occur periodically. During epidemics, onset is rapid; many animals are affected within days or 2–3 wk. Bovine ephemeral fever is most prevalent in the wet season in the tropics and in summer to early autumn in the subtropics or temperate regions (when conditions favor multiplication of biting insects); it disappears abruptly in winter. Virus spread appears to be limited by latitude rather than topography or availability of susceptible hosts. Morbidity may be as high as 80%; overall mortality is usually 1%–2%, although it can be higher in lactating cows, bulls in good condition, and fat steers (10%–30%).

Clinical findings

Signs, which occur suddenly and vary in severity, can include biphasic to polyphasic fever (40°–42°C [104°–107.6° F), shivering, inappetence, lacrimation, serous nasal discharge, drooling, increased heart rate, tachypnea or dyspnea, atony of forestomachs, depression, stiffness and lameness, and a sudden decrease in milk yield.

Affected cattle may become recumbent and paralyzed for 8 hr to >1 wk. After recovery, milk production often fails to return to normal levels until the next lactation. Abortion, with total loss of the season's lactation, occurs in about 5% of cows pregnant for 8–9 months. The virus does not appear to cross the placenta or affect the fertility of the cow. Bulls, heavy cattle, and high-lactating dairy cows are the most severely affected, but spontaneous recovery usually occurs within a few days. More insidious losses may result from decreased muscle mass and lowered fertility in bulls.

Lesions

The most common lesions include polyserositis affecting pleural, pericardial, and peritoneal surfaces; serofibrinous polysynovitis, polyarthritis, polytendinitis, and cellulitis; and focal necrosis of skeletal muscles. Generalized edema of lymph nodes and lungs, as well as atelectasis, also may be present.

Treatment and control

Complete rest is the most effective treatment, and recovering animals should not be stressed or worked because relapse is likely. Anti-inflammatory drugs given early and in repeated doses for 2–3 days are effective. Oral dosing should be avoided unless the swallowing reflex is functional. Signs of hypocalcemia are treated as for milk fever. Antibiotic treatment to control secondary infection and rehydration with isotonic fluids may be warranted.

RINDERPEST

Rinderpest is the most destructive of the virus diseases of cloven-footed animals, such as cattle, buffaloes, sheep, goats, pigs and wild ruminants. The virus is found notable in the saliva, discharge from eyes and nostrils, and in the urine and faeces. It is present in the circulating blood during the febrile stage and is later concentrated in different organs, especially in the spleen, lymph nodes and liver. Outside the animal body, the virus is rapidly destroyed by direct sunlight and disinfectants. Cold preserves the virus. The virus is usually spread by contaminated feed and water. Rise in temperature up to 104 – 107 degree F. Lacrimation and redness of eye. Foul odour from mouth. Discrete necrotic foci develop in the buccal mucosa, inside lip, and on the tongue. Bloody mucoid diarrhea is noticed

Treatment

Symptomatic treatment can be help early cure of the animals. Consult with veterinary doctor

MASTITIS

Mastitis, or inflammation of the mammary gland, is the most common and the most expensive disease of dairy cattle throughout most of the world. Although stress and physical injuries may cause inflammation of the gland, infection by invading bacteria or other microorganisms (fungi, yeasts and possibly viruses) is the primary cause of mastitis. Infections begin when microorganisms penetrate the teat canal and multiply in the mammary gland.

Treatment success depends on the nature of the aetiological agent involved, the severity of the disease and the extent of fibrosis. complete recovery with freedom from bacterial infection can be obtained in cases of recent infection and in those where fibrosis has taken place only to a small extent.

Such drugs as acriflavine, gramicidin and tyrothricin have now ceased to be in use, and have given place to the more effective drugs, such as sulphonamides, penicillin and streptomycin.

FOOTROT

Foot rot is a common cause of lameness in cattle and occurs most frequently when cattle on pasture are forced to walk through mud to obtain water and feed. However, it may occur among cattle in paddocks as well, under apparently excellent conditions. Foot rot is caused when a cut or scratch in the skin allows infection to penetrate between the claws or around the top of the hoof. Individual cases should be kept in a dry place and treated promptly with medication as directed by a veterinarian. If the disease becomes a herd problem a foot bath containing a 5% solution of copper sulphate placed where cattle are forced to walk though it once or twice a day will help to reduce the number of new infections. In addition, drain mud holes and cement areas around the water troughs where cattle are likely to pick up the infection. Keep pens and areas where cattle gather as clean as possible. Proper nutrition regarding protein, minerals and vitamins will maximize hoof health.

BOVINE RHINOTRACHEITIS

Infectious bovine rhinotracheitis (ibr) is a highly contagious, infectious respiratory disease that is caused by bovine herpesvirus-1 (bhv-1). It can affect young and older cattle. In addition to causing respiratory disease, this virus can cause conjunctivitis, abortions, encephalitis, and generalised systemic infections. Ibr is characterized by acute inflammation of the upper respiratory tract.

Treatment

There is no direct treatment for viral diseases. Infected animals should be isolated from the rest of the herd and treated with anti-inflammatory drugs and antibiotics for secondary infections if necessary. Carrier cattle should be identified and removed from the herd.

Prevention

Control of the disease is based on the use of vaccines.

PIGLET DIARRHEA OR SCOUR

Of all the diseases in the sucking piglet, diarrhea is the most common and probably the most important. In some outbreaks, it is responsible for high morbidity and mortality. The main bacterial causes are E. Coli and Clostridia and the main parasite is Coccidia.

Clinical signs

Scour in the piglet can occur at any age during sucking but there are often two peak periods, before 5 days and between 7 and 14 days.

Acute disease

The only sign may be a perfectly good pig found dead. Post-mortem examinations show severe acute enteritis, so sudden that there may be no evidence of scour externally. Clinically affected piglets huddle together shivering or lie in a corner. The skin around the rectum and tail will be wet. Look around the pen for evidence of a watery to salad cream consistency scour. In many cases, there is a distinctive smell. As the diarrhea progresses the piglet becomes dehydrated, with sunken eyes and a thick leathery skin. The scour often sticks to the skin of other piglets giving them an orange to white color.

Prior to death piglets may be found on their sides paddling and frothing at the mouth.

Sub-acute disease

The symptoms are similar but the effects on the piglet are less dramatic, more prolonged and mortality tends to be lower. This type of scour is often seen between 7 to 14 days of age manifest by a watery to thin salad cream consistency diarrhea, often white to yellow in color.

Treatment in severe outbreaks of E. Coli disease the sows feed can be top dressed with the appropriate antibiotic daily, from entry into the farrowing house and for up to 14 days post-farrowing. This can be effective in reducing bacterial output in the sows faeces. observe litters for the presence of diarrhea both night and morning. study the history of the disease on your farm. Is it sporadic, in one piglet in a litter, or total litters? in the light of the history either treat the individual pig or on the first signs of disease treat the whole litter. if a litter is badly scoured dose night and morning for a minimum of two days. assess the response to treatment. If there is no change within 12 hours, then change to another medicine as advised by your veterinarian. always treat piglets less than 7 days of age by mouth. for older pigs where the disease is less acute injections are equally effective and easier to administer. provide electrolytes in drinkers. These prevent dehydration and maintain body electrolyte balances. cover the pen, the creep area and where the pigs defecate with straw, shredded paper, shavings or sawdust. provide an additional lamp to provide an extra source of heat. use binding agents such as chalk, kaolin or activated attapulgite to absorb toxins from the gut.

Management control and prevention adopt procedures to prevent the spread of the scour - disinfect boots between pens, use a disposable plastic apron when dosing piglets to prevent heavy contamination of clothing, wash hands after handling a scoured litter, disinfect brushes and shovels between pen. ensure that farrowing houses are only used on an all-in all-out basis with a pressure wash and disinfection between each batch. farrowing pens must be dry before the house is repopulated. Remember that moisture, warmth, waste food and faeces are ideal for bacterial multiplication. pen floors should be well maintained. Poor pen hygiene associated with bad drainage predisposes to scour. look carefully at the part of the pen floor where there are piglet faeces. Is this poorly drained? Do large wet patches develop? If so cover them with extra bedding daily and remove. This is a most important aspect of control.

Check nipple drinkers and feeding troughs for leakages. Cnsure that faeces are removed daily from behind the sow from the day she enters the farrowing crates until at least 7 days post-farrowing if the floors are slatted. Also remove faeces daily throughout lactation if they are solid concrete. maintain creep environments that are always warm and comfortable. Fluctuating temperatures are a major trigger factor to scour particularly from 7 to 14 days of age.

Consider vaccinating against E. Coli (make sure first that this is the cause of the problem however). E. Coli vaccines only protect the piglet for the first 5 to 7 days of age. assess the environment of all the farrowing house. Poor environments allow heavy bacterial multiplication and a much higher bacterial challenge is likely to break down the colostral immunity.

Check the sow's health. Animals affected with enteric or respiratory disease, lameness or mastitis predispose the litter to scour.

Where farrowing house floors are very poor, pitted and difficult to clean, brush them over with lime wash containing a phenolic disinfectant.

Colostrum management : it is vital that the piglet receives the maximum amount of colostrum within the first 12 hours of birth. High levels of antibody are only absorbed during this period. Factors such as poor teat access, poor crate design, and particularly the development of agalactia in the sow, associated with udder oedema, reduce intake.

PPR (GOAT PLAGUE)

PPR (Peste des petits ruminants) is a most important viral disease of goat capable of heavy mortality and commonly called as goat plague.

Etiology

The causative virus was first thought to be an aberrant strain of rinderpest virus that had lost its ability to infect cattle. Later molecular studies showed that it was distinct from, but closely related to, rinderpest virus.

Clinical signs

The clinical sign of PPR in goats is often fulminating and fatal although apparent infection occurs in endemic areas. Incubation period may range from 2-6 days in field conditions. In acute form, there is sudden onset of fever with rectal temperature of at least 40°- 41°C. The affected goats show dullness, sneezing, serous discharge from the eyes and nostrils. During this stage farmers often think that the animal has developed cold exposure and may attempt to provide protection for cold. In the process goats, may be congregated and accentuate the process of transmission. After 2-3 days, discrete lesions develop in the mouth and extend over the entire oral mucosa, forming diphtheric plaques.

During this stage profound halitosis (foul smell) is easily appreciable and the animal is unable to eat due to sore mouth and swollen lips. Latter ocular discharge becomes mucopurulent and the exudate dries up, matting the eyelids and partially occluding the nostrils. Diarrhea develops 3-4 days after the fever and is profuse and faeces may be mucoid or bloody depending upon the damage. Dyspnea and coughing occur later due to secondary pneumonia. Death occurs within one week of the onset of the illness.

Treatment and control

No specific treatment is recommended for ppr being viral disease. However, mortality rates can be reduced by the use of drugs that control the bacterial and parasitic complications. Specifically, oxytetracycline and chlortetracycline are recommended to prevent secondary pulmonary infections. Lesions around the eyes, nostrils and mouth should be cleaned twice daily with sterile cotton swab.

Our experience indicates that fluid therapy and anti-microbial such as enrofloxacin or ceftiofur on recommended doses along with mouth wash with 5% boro-glycerine can be of benefit in reducing the mortality during outbreak of ppr in goats. Health workers should inspect first the unaffected goats followed by treatment of affected goats. Immediate isolation of affected goats from clinically healthy goats is most importance measure in controlling the spread of infection. Nutritious soft, moist, palatable diet should be given to the affected goats. Provide parenteral energy infusion in anorectic goats along with appetizers.

Immediately measures should be taken for notification of disease to nearest government veterinary hospital. Carcasses of affected goats should be burned or buried. Proper disposal of contact fomites, decontamination is must. Vaccination is the most effective way to control ppr.

BOVINE BABESIOSIS (TICK FEVER)

Cause

Bovine babesiosis (bb) is a tick-borne disease of cattle. Transmission of b bovis takes place when engorging adult female ticks pick up the infection. They pass it on to their progeny via their eggs. Larvae (or seed ticks) then pass it on in turn when feeding on another animal. B bigemina is also passed from one generation of ticks to the next. Engorging adult ticks pick up the infection and nymphal and adult stages (not larval stages) of the next generation pass it on to other cattle.

Morbidity and mortality vary greatly and are influenced by prevailing treatments employed in an area, previous exposure to a species/strain of parasite, and vaccination status. In endemic areas, cattle become infected at a young age and develop a long-term immunity. However, outbreaks can occur in these endemic areas if exposure to ticks by young animals is interrupted or immuno-naïve cattle are introduced. The introduction of babesia infected ticks into previously tick-free areas may also lead to outbreaks of disease.

Symptoms:

High fever neurologic signs such as incoordination, teeth grinding and mania. Some cattle may be found on the ground with the involuntary movements of the legs. When the nervous symptoms of cerebral babesiosis develop, the outcome is almost always fatal. dark colored urine anorexia animals likely to separate from herd, be weak, depressed and reluctant to move n b. Bigemina parasitaemia often exceeds 10 per cent and may be as high as 30 per cent.

Clinical symptoms for babesia divergens are similar to b. Bigemina infections. The survivors may be weak and in reduced condition, although they usually recover fully. Subacute infections, with less apparent clinical signs, are also seen.

Treatment

Mild cases may recover without treatment. Sick animals can be treated with an antiparasitic drug. Treatment is most likely to be successful if the disease is diagnosed early; it may fail if the animal has been weakened by anemia. Imidocarb has been reported to protect animals from disease but immunity can develop. There are also concerns with regard to residues in milk and meat. In some cases blood transfusions and other supportive therapy should be considered.

Prevention

Effective control of tick fevers has been achieved by a combination of measures directed at both the disease and the tick vector. Tick control by acaracide dipping is widely used in endemic areas. Dipping may be done as frequently as every 4-6 weeks in heavily infested areas. The occurrence of resistance of ticks, chemical residues in cattle and environmental concerns over the continued use of insecticides has led to use of integrated strategies for tick control.

Babesiosis vaccines are readily available and are highly effective. Anti-tick vaccines are also available in some countries and can be used as part of an integrated program for the control of ticks. Babesiosis can be eradicated by eliminating the host tick(s). In the us, this was accomplished by treating all cattle every two to three weeks with acaricides. In countries where eradication is not feasible, tick control can reduce the incidence of disease.

Treatment for control of tick (ethnovet)

Mix common salt and few camphor in castor oil or neem oil and apply over the affected area. Whole plant extract of ghaner (lantana camara) should be diluted with the urine of cattle and apply externally. Boil 250 gm of tobacco in 2 litres of water and add 5 litres of water and sprayed over the body of 10-20 animals.

Theileriosis

Theileriases are a group of tickborne diseases caused by theileria spp. Both theileria and babesia are members of the suborder piroplasmorina. Although babesia are primarily parasites of rbcs, theileria use, successively, wbcs and rbcs for completion of their life cycle in mammalian hosts. The infective sporozoite stage of the parasite is transmitted in the saliva of infected ticks as they feed. Sporozoites invade leukocytes and, within a few days, develop to schizonts. In the most pathogenic species of theileria (eg, t parva and t annulata), parasite multiplication occurs predominantly within the host wbcs, whereas less pathogenic species multiply mainly in rbcs.

Development of the schizont stage of pathogenic theileria causes the host wbc to divide; at each cell division, the parasite also divides. Mortality in such stock is relatively low, but introduced cattle are particularly vulnerable. Unlike in babesiosis, in theileriasis there is no evidence of increased resistance in calves <6 mo old.

East coast fever

East coast fever, an acute disease of cattle, is usually characterized by high fever, swelling of the lymph nodes, dyspnea, and high mortality. Caused by theileria parva, and transmitted by the tick vector rhipicephalus appendiculatus, it is a serious problem in east and southern africa.

Etiology and transmission

The african buffalo (syncerus caffer) is an important wildlife reservoir of t parva, but infection is asymptomatic in buffalo. T parva transmitted by ticks from either cattle or buffalo cause severe disease in cattle, but buffalo-derived parasites differentiate poorly to merozoites in cattle and generally are not transmitted by ticks. Hence, buffalo t parva are maintained as a separate population. Buffalo t parvawere previously considered a separate subspecies (t parva lawrencei), but dna typing indicate that the cattle and buffalo parasites are a single species. T parva is usually highly pathogenic, causing high levels of mortality, although some less pathogenic isolates have been identified.

Pathogenesis, clinical findings, and diagnosis

T parva sporozoites are injected into cattle by infected vector ticks. An occult phase of 5–10 days follows before infected lymphocytes can be detected in giemsa-stained smears of cells aspirated from the local draining lymph node. Subsequently, the number of parasitized cells increases rapidly throughout the lymphoid system, and from about day 14 onward, cells undergoing merogony are observed. This is associated with widespread lymphocytolysis, marked lymphoid depletion, and leukopenia. Piroplasms in rbcs infected by the resultant merozoites assume various forms, but typically they are small and rod-shaped or oval.

Clinical signs vary according to the level of challenge, and they range from in apparent or mild to severe and fatal. Typically, fever occurs 7–10 days after parasites are introduced by feeding ticks, continues throughout the course of infection, and may be >106°f (41°c). Lymph node swelling becomes pronounced and generalized. Lymphoblasts in giemsa-stained smears of needle aspirates from lymph nodes contain multinuclear schizonts. Anorexia develops, and the animal rapidly loses condition; lacrimation and nasal discharge may occur. Terminally, dyspnea is common.

Just before death, a sharp decrease in body temperature is usual, and pulmonary exudate pours from the nostrils. Death usually occurs 18–24 days after infection. The most striking postmortem lesions are lymph node enlargement and massive pulmonary edema and hyperemia. Hemorrhages are common on the serosal and mucosal surfaces of many organs, sometimes together with obvious areas of necrosis in the lymph nodes and thymus. Anemia is not a major diagnostic sign (as it is in babesiosis) because there is minimal division of the parasites in rbcs, and thus no massive destruction of them.

Animals that recover are immune to subsequent challenge with the same strains but may be susceptible to some heterologous strains. Most recovered or immunized animals remain carriers of the infection.

Treatment and control

Treatment with parvaquone and its derivative buparvaquone is highly effective when administered in the early stages of clinical disease but is less effective in the advanced stages, in which there is extensive destruction of lymphoid and hematopoietic tissues. Immunization of cattle against t parva using an infection-

and-treatment procedure is practical and continues to gain acceptance in some regions.

The components for this procedure are a cryopreserved sporozoite stabilate of the appropriate strain(s) oftheileria derived from infected ticks and a single dose of long-acting oxytetracycline given simultaneously; although oxytetracycline has little therapeutic effect when administered after development of disease, it inhibits development of the parasite when given at the outset of infection. Cattle should be immunized 3–4 wk before being allowed on infected pasture. Parasitized bovine cells containing the schizont stage of t parva and t annulata can be cultivated in vitro as continuously growing cell lines. In the case of t annulata, cattle can be infected with a few thousand cultured cells. Attenuated strains produced by serial passage of such cultures form the basis of live vaccines used in several countries, including israel, iran, india, and the former ussr.

Incidence of east coast fever can be reduced by rigid tick control, but this is not feasible in many areas because of cost and the high frequency of acaricidal treatment required.

RINGWORM

This is the most common infectious skin disease affecting beef cattle. It is caused by a fungus, and is transmissible to man. Typically, the disease appears as crusty grey patches usually in the region of the head and neck and particularly around the eyes.

As a first step in controlling the disease, it is recommended that, whenever possible, affected animals should be segregated and their pens or stalls cleaned and disinfected. Clean cattle which have been in contact with the disease should be watched closely for the appearance of lesions and treated promptly. Proper nutrition, particularly high levels of vitamin a, copper and zinc while not a cure, will help to raise the resistance of the animal and in so doing offer some measure of control. Contact your vet and or feed store for products to treat this disease. Using a wormer like ivomec will kill lice and help prevent cattle from scratching causing skin damage and a place for the fungus to enter.

MILK FEVER

Milk fever, also known as Parturient hypocalcaemia and parturient paresis, is a disease which has assumed considerable importance with the development of heavy milking cows. Decrease in the levels of ionized calcium in tissue fluids is basically the cause of the disease. In all adult cows, there is a fall in serum-calcium level with the onset of lactation at calving. The disease usually occurs in 5 to 10-year-old cows, and is chiefly caused by a sudden decrease in blood-calcium level, generally within 48 hours after calving.

Symptoms

In classical cases, hypocalcaemia is the cause of clinical symptoms. Hypophosphataemia and variations in the concentration of serum-magnesium may play some subsidiary role.

The clinical symptoms develop usually in one to three days after calving. They are characterized by loss of appetite, constipation and restlessness, but there is no rise in temperature.

CALF SCOUR

Calves may develop scours due to bacterial or virus infections. Scours is known as "calf scours" or neonatal calf diarrhea. The primary causes of scours include: Rota virus, Corona virus, Cryptosporidium parvum, Salmonella and Escherichia coli.

Determine if treatment is required. Calves that are moving around in the pasture, with their tails up, probably do not need treatment. Check to see if the diarrhea is yellow or white. If this is the case, treatment is probably not needed.

Determine if the calf is looking listless. Calves that are lethargic or not participating much in the playful activities with other calves are a red flag to pay attention to. Calves that are also losing condition are also cause for alarm.

Check to see if the calf is dehydrated. You can check for dehydration by pulling on the calf's neck skin. If the skin "tents" this is a sign of dehydration.

Determine the calf's body temperature. A normal body temperature ranges from 100.5 °f (38.1 °c) to 102.5 °f (39.2 °c). Anything outside of this range is a sign for treatment.

Separate the sick calf or calves from the healthy herd. You'll want to do this to avoid spreading the disease further.

Administer fluids using your veterinarian-approved electrolyte solution. You may need to inject the fluids via iv or orally.

Follow appropriate nursing care protocol using your vet's guidelines. This may include providing shelter, feed and a warm place to sleep.

3

VIRAL DISEASES OF ANIMAL

Global change refers to environmental changes arising from human activities affecting the fundamental mechanisms operating in the biosphere. This paper discusses the changes observed in recent times with regard to some important arboviral (arthropod-borne viral) diseases of animals, and the role global change could have played in these variations. Two of the most important arboviral diseases of animals, bluetongue (BT) and West Nile fever/encephalitis (WNF), have been selected as models. In both cases, in the last 15 years an important leap forward has been observed, which has lead to considering them emerging diseases in different parts of the world. BT, affecting domestic ruminants, has recently afflicted livestock in Europe in an unprecedented epizootic, causing enormous economic losses.

WNF affects wildlife (birds), domestic animals (equines), and humans, thus, beyond the economic consequences of its occurrence, as a zoonotic disease, it poses an important public health threat. West Nile virus (WNV) has expanded in the last 12 years worldwide, and particularly in the Americas, where it first occurred in 1999, extending throughout the Americas relentlessly since then, causing a severe epidemic of disastrous consequences for public health, wildlife, and livestock. In Europe, WNV is known long time ago, but it is since the last years of the twentieth century that its incidence has risen substantially.

Infectious disease can be viewed as a play involving at least two characters: the pathogen and the host. While both roles can be represented by a great variety of performers, pathogens exhibit by far the highest variety and complexity. This review is about viral infections in animals. It aims first to give an idea of the enormous complexity and diversity of the existing infectious agents, emphasizing their extraordinary capacity for change and adaptation, which eventually leads to the emergence of new infectious diseases. Secondly, it focuses on the influence of the environment in this process, and on how environmental (including climate) changes occurring in recent times, have precise effects on the emergence and evolution of infectious diseases, some of which will be illustrated with specific examples.

IDENTIFICATION OF VIRAL DISEASES

Identification and control of viruses that causes diseases in human, domestic and wild animals represent continuous challenges to medical and veterinary sciences. From centuries many diagnostic techniques have been tried to identify individual infectious agents by their clinical signs and symptoms produced in the susceptible host. But it is not always possible to identify or differentiate the disease on the basis of its signs and symptoms and thus requires other advanced diagnostic procedures to clearly identify and confirm the pathogens. Rapid detection of infectious agents in animals as well as in environment is essential for effective control of infectious diseases.

There are various approaches for detection of viral diseases which includes conventional as well as modern approaches. Several conventional techniques such as isolation in cell culture, serology and histological identification are traditionally used to identify viral pathogens. However, the latest molecular diagnostic technologies are more advantageous as they are offer more sensitive, less time consuming, high throughput result. Other novel technologies such as polymerase chain reaction (PCR), Reverse Transcriptase PCR (RT–PCR), Real Time–PCR, DNA probe, nucleic acid sequencing etc. provide a thorough understanding of accurate diagnosis and discrimination of present and emerging diseases.

The aim of these new tools is to detect the presence of pathogen before the onset of disease. Here we have discussed the advancement of diagnostic tools and their applications for important animal viruses of domestic and wild animals.

VIRAL DISEASES OF ANIMALS

Complexity and Diversity of Animal Viruses

To illustrate the enormous complexity of animal viruses, consider the following example: take any animal species, e.g., bovine. There are five known species of herpesviruses that could infect bovines specifically. Similarly, there are nine different known equine herpesviruses, eight human herpesviruses, and so on and so forth (Pellet and Roizman, 2007). Bearing in mind that there are approximately 5,400 different species of mammals (Wilson and Reeder, 2005) and that most of them have yielded at least one, most frequently several distinct herpesviruses on examination, the number of existing mammalian herpesvirus species would be huge, probably in the range of thousands. But there are also herpesviruses specific for birds, for reptiles, for amphibians, etc., so the above number would be increased consequently with the number of other vertebrate species (for the moment viruses of invertebrates, a world largely to be discovered, will not be considered).

Likewise, let's bear in mind that there are other taxonomic families of viruses besides the Herpesviridae family, like the Poxviridae (e.g., smallpox and myxomatosis), Flaviviridae (e.g., yellow fever and dengue), Orthomyxoviridae (e.g., influenza), Picornaviridae (e.g., foot-and-mouth disease, polio, and hepatitis A), Reoviridae (e.g., BT and African horse sickness), etc., and for each family of viruses one can reason about in the same way. In light of this example, a first conclusion is that we only know a fraction of the viral pathogens that actually exist.

To these, we must add the non-pathogenic viruses that circulate silently, which obviously are less known, and which probably exist in far greater numbers and variety than their pathogenic counterparts. The complexity of viruses of plants, bacteria, fungi, and parasites is not lower than that of animal viruses. This gives a rough idea of the real complexity of the virus world in which only a small part is known.

The Genesis of Infectious Diseases

Despite the distress and alarm that the emergence of a new infectious disease causes, the fact is that organisms change over time in adaptive processes that determine their evolution, and this implies the emergence of new infectious diseases and the disappearance or change of the existing ones. These processes are therefore normal and expected to occur with some frequency, as a result of the high generation rates and enormous capacity for mutation and adaptation exhibited by microorganisms in general, and viruses in particular.

The term emerging infectious diseases applies to those diseases in which any of the following situations is applicable: (1) a known infection spreading to a new geographic area or population, (2) a new infection that occurs as a result of evolution or change of an existing pathogen, and (3) a previously unknown disease or pathogen that is first diagnosed (Brown, 2004; OIE, 2011). Factors determining the emergence of an infectious disease are largely unknown, although they are related to the adaptation of infectious agents to new species (Lederberg, 1997) and/or to the concomitant appearance of changes in the environment that offer new opportunities for pathogens to thrive (Smolinski et al., 2003). With regard to adaptation to new species, first it is worth reminding that every virus has its own "host range," that is, the variety of species susceptible to be infected by a particular virus. There are, on one hand, viruses with a broad range of hosts, for example, West Nile virus (WNV), that is capable of infecting hundreds of species of birds, as well as many species of mammals, reptiles, and amphibians (McLean et al., 2002), and, on the other hand, viruses with a narrower range of hosts such as, for instance, classical swine fever virus, which only infects suidae (pigs and wild boar; OIE, 2008), or the aforementioned herpesviruses, each of which usually infect specifically one or a few related species (Pellet and Roizman, 2007).

Two cases can be distinguished with regard to pathogen adaptation to new host species: in the first, the pathogen completes its cycle and survives normally in a given species which acts as reservoir host, but occasionally infects other(s), causing disease in these "incidental hosts." This occurs in a number of zoonoses. A good example is given by highly pathogenic avian influenza virus subtype H5N1 of Asian origin (H5N1 HPAIV), now present on three continents. Different wild birds act as reservoir hosts for this virus, which affects severely different species of domestic poultry, causing enormous losses to the poultry industry (Alexander, 2000).

However, besides birds, it occasionally affects some species of mammals, including humans (causing a zoonosis), cats, and mustelids like ferrets and martens, in which it is extremely pathogenic. Nevertheless, transmission of H5N1 HPAIV between individuals in mammal species is not effective at all (Imai and Kawaoka, 2012), a fact that luckily prevents by now a possible pandemic of disastrous consequences for humankind. In the second case, the adaptation to new species is more "genuine" in the sense that the pathogen has indeed crossed the "species barrier," that is, has established a complete cycle of transmission in a new species. It can be assumed that all viruses currently known went through a process like this in their past evolution to adapt to the species which are their current natural hosts. Influenza viruses themselves were adapted to their avian hosts in remote times.

Some of them gave rise, through analogous adaptation processes, to the current swine, equine, and human influenza viruses, which are able to complete an infectious cycle in these species, independently of birds (Suarez, 2000). Among the viruses that have undergone a process of adaptation to a new species, it is worth reminding the case of swine vesicular disease virus, which causes an economically important disease in pigs (Escribano-Romero et al., 2000). This virus is closely related genetically to Coxsackie virus B5, a human enterovirus, and current evidence suggests that it arose as a result of a relatively recent adaptation from human to swine (Jimenez-Clavero et al., 2005a), providing a good example that adaptations and crossing the species barrier can go either to or from humans, and that the human being is just one more species in this regard.

Other viruses that have recently emerged to affect new species are several variants of the genus Henipavirus, infecting bats in Southeast Asia (Nipah virus) and Oceania (Hendra virus; Eaton et al., 2006). In recent years an increase in fatal cases by these viruses in humans and livestock (pigs and horses) has been observed, but an inefficient transmission between individuals of these species has prevented further spread in the human and/or livestock populations. However, in a recent outbreak of Nipah virus occurring in Bangladesh it appeared that an efficient transmission occurred between humans (Gurley et al., 2007).

Finally, it is important to note that there is probably a species barrier crossing after every first diagnosis of a viral emerging disease, even though the natural reservoir might remain unknown. In 2003, the first cases of what appeared to be a new disease entity with fatal consequences for affected humans, were diagnosed in Hong Kong. It received the name of severe acute respiratory syndrome (SARS) and caused a great social concern worldwide, particularly when clusters of cases were detected in up to a dozen countries in three continents around the world, although a rapid response prevented further spread. The causative agent of this disease, named SARS virus, was identified during earlier investigations (Drosten et al., 2003; Kuiken et al., 2003), but its natural host remained unknown until Rhinolophidae (horseshoe bats) species were identified as reservoir hosts (Li et al., 2005; Cui et al., 2007).

INFLUENCE OF THE ENVIRONMENT IN INFECTIOUS DISEASES

The environment of viruses is constituted essentially by the cells they infect. Viruses must parasitize cells to ensure their propagation. Outside cells, viruses lack of activity, although to allow transmission to another individual they need to survive outside the cell. The way a virus survives in the environment is the result of an adaptation that largely determines how it is transmitted. Viruses can be transmitted through direct contact between individuals, or through other ways such as, for instance air, water, feces, body fluids, food, or fomites. In some cases, transmission needs the participation of other living organisms, so-called vectors, in which the virus is equally propagated. These vectors are often blood-sucking insects that inoculate the infectious pathogen through their bites, thus spreading the infection in a population. A virus may preferably use a single transmission route, but viruses using more than one way of transmission are frequent.

The progress of a viral infection in a population is a multifactorial process, depending on a range of both biotic and abiotic factors. These factors and their influence on the development and transmission of the infection at the population level constitute the eco-epidemiology of an infectious disease. There is a strong relationship between the route of transmission and the eco-epidemiology of a given infectious disease. For example, the spread of infections that are transmitted by direct contact largely depends on the population density, which determines the distance between infected and susceptible individuals. For airborne infections, temperature, humidity, and wind can have a significant influence on the progress of the epidemic. Waterborne and foodborne viruses are usually highly resistant to adverse environmental conditions. A particular case of this type of transmission is represented by the fecal–oral route (often water, food, or objects are contaminated by fecal waste).

Viruses that are transmitted through this route have a characteristic resistance to low pH, allowing them to pass through the animal's digestive tract, overcoming the natural barrier that represents the acid secretion of gastric parietal cells. Often, these viruses produce diarrhea, thus being shed in large amounts and returning to the environment, where they can remain infectious for a variable period (up to several months in some instances), depending mainly on environmental temperature (the colder the longer), but also on the presence of salts, organic matter, moisture, solar radiation, etc., until they reach another host and begin the infectious cycle again (Jimenez-Clavero et al., 2005b).

In the case of vector-borne diseases, in addition to pathogen and host, infectious cycle requires a third player: the vector. This fact results in a more complex eco-epidemiology. Habitats of arthropod vectors depend on a number of environmental conditions, including range of temperature, humidity, water availability, etc. Arboviruses (i.e., viruses transmitted by arthropod vectors), are distributed necessarily in areas where populations of competent vectors (i.e., vectors which are able to spread the disease to new hosts) are abundant enough. Each arthropod vector species occupies a particular ecological niche within specific environmental conditions, so that its distribution can be greatly affected by changes in temperature, rainfall, humidity, plant coverage, etc. (Randolph and Rogers, 2010; Weaver and Reisen, 2010).

In summary, environmental variations constitute an important factor in the occurrence and spread of infectious diseases in general. This is particularly important for arthropod-borne diseases, where environmental changes can influence both their geographic range and the risk of introduction (Sutherst, 2004; Gould and Higgs, 2009; Tabachnick, 2010).

Emerging Infectious Diseases and Global Change

The Planet Earth is in continuous change. Environmental changes are driven by both abiotic and biotic factors. Of these, impact caused by humans has gained importance as the human population has risen, from the local level to reach a global scale (Meadows et al., 1972). Global change is defined as the impact of human activity on the fundamental mechanisms operating in the biosphere (Duarte, 2006). Global change comprises not only impacts on climate, but also on the water cycle, land use, biodiversity loss, invasion of alien species into new territories, introduction of new chemicals in Nature, etc. The influence of global change on emerging infectious diseases has been the subject of different revisions .

As noted above, environmental changes largely determine the evolution of many emerging infectious diseases, most notably arthropod-borne ones (see reviews by Sutherst, 2004; Gould and Higgs, 2009; Tabachnick, 2010), Therefore, climate changes will impact necessarily on these diseases (Rosenthal, 2009). However, there are many other driving factors – not only those related to climate changes – that have or have had a powerful influence on the occurrence or change of many infectious diseases. Table 1 shows some of these drivers, identified as relevant in the emergence and re-emergence of infectious diseases in a recent study (Woolhouse and Gowtage-Sequeria, 2005).

For instance, intercontinental transport by air makes it possible to move persons and goods thousands of miles away within few hours. For a person who has just acquired an infection it is possible to arrive to destination even within the incubation period, and develop (and transmit) the disease upon arrival, resulting in an "imported disease." This not only applies to people, but occurs similarly as a result of trade of live animals and their products, an important economic activity worldwide, which is subjected to strict regulations (OIE, 2011) that must be implemented in coordination with all countries, precisely to prevent the spread of infectious diseases harmful to livestock, which in this context are called "exotic" or "transboundary" diseases.

However, these rules have been insufficient to halt the spread of many infectious animal diseases of tremendous economic impact, partly because control and surveillance systems do not always work effectively. As an example of what an effective control can achieve, it is worth to mention what constituted the first detection in Europe of highly pathogenic avian influenza virus H5N1 from Asia, which occurred in 2004 at a border checkpoint in the airport of Brussels (Belgium). Customs officers detected in the luggage of a traveler from Thailand two mountain hawk eagles (Spizaetus nipalensis) alive, apparently brought as a gift.

The two birds were found to be infected with the virus (Van Borm et al., 2005). This case also reminds that, in addition to the above, illegal trafficking of animals, including exotic species, must also be considered as a relevant factor involved in importation of transboundary diseases. A good example was given in 2003 by the occurrence of outbreaks of monkeypox in humans in the U.S., originated as a result of illegal import of infected exotic rodents from Ghana (Guarner et al., 2004). The disease has reached local populations of rodents (prairie dogs, Cynomys spp.).

Regarding arboviral diseases, the impact of trade and transport on the distribution of vectors is reflected by the global expansion of the tiger mosquito (Aedes albopictus), associated with trade in used tires (Reiter and Sprenger, 1987) or in Dracaena plants ("lucky bamboo"; Madon et al., 2002). Rain causes small pools of water inside the tires stored outdoors. This constitutes an excellent breeding habitat for this mosquito, because it mimics the hollow trunks of rainforest trees that are its natural habitat. Through transport of used tires containing A. albopictus eggs, this mosquito has reached a worldwide distribution. This mosquito is a competent vector of many pathogens, including dengue, yellow fever, chikungunya, Venezuelan equine encephalitis, and WNF (Paupy et al., 2009; Weaver and Reisen, 2010).

Other factors derived from human activities linked to the emergence of infectious diseases are those related to land use: a growing human population is demanding more resources necessary for its supply and welfare, particularly food, but also energy, raw materials, water, urban land, etc. New agricultural plantations and livestock grazing plots on deforested lands promote the emergence and/or spread of infectious diseases hitherto confined to the forest habitat (Smolinski et al., 2003; Woolhouse and Gowtage-Sequeria, 2005).

Deforestation and biodiversity loss are often cited as causes of increased incidence of emerging infections in certain regions. An example can be found in the emergence of Junin virus in Argentina (Charrel and de Lamballerie, 2003). Changes in water use are, nonetheless, one of the factors most clearly related to the modification of eco-epidemiological patterns accompanying emerging infectious diseases. New irrigation plans, construction of dams, etc. have a direct effect on the abundance of competent vectors for transmission of various arboviruses.

The new irrigation in northwestern Australia seem to be the main cause of the recent expansion of Murray Valley fever, an emerging zoonotic arbovirus in that country (Mackenzie et al., 2004). The construction of dams may have caused an important impact on the recent emergence of Rift Valley fever in East Africa, causing highly virulent outbreaks for both man and livestock, by allowing a dramatic increase in mosquito populations involved in transmission (Martin et al., 2008). Healthcare is another area of human activity that, somehow paradoxically, is linked in many instances to infectious disease emergence. Iatrogenic transmission of infectious diseases, through transfusions, transplants, and other medical interventions, has undoubtedly had an effect on the expansion of certain pathogens such as hepatitis C virus. Fortunately, current medical practice has reduced this risk very significantly. The administration of biologicals derived from animals or animal cell cultures, such as vaccines and therapeutic products, can also act as a vehicle for the transmission of pathogens.

Despite strict controls on each batch of vaccine for the presence of certain pathogens, there have been cases of transmission of adventitious viruses in contaminated vaccines. One of the best known cases of this kind is represented by bovine viral diarrhea virus (BVDV), a pestivirus affecting cattle. In the manufacturing process of many vaccines, cell cultures are widely employed, and fetal calf serum is used as an additive common to cell culture media. BVDV is highly prevalent in cattle and certain variants of the virus remain unnoticed in cell cultures.

In some cases, batches of fetal calf serum from BVDV-infected animals were used inadvertently, resulting in viral contamination of vaccines. Some of them (those addressed to the bovine, such as vaccine against bovine herpesvirus type 1) eventually resulted in outbreaks of viral diarrhea in vaccinated cattle, which alerted for the presence of the virus (Makoschey et al., 2003).

Currently, regulatory agencies, such as the European Medicines Agency, have modified their safety requirements to control specifically the risk of contamination by animal pestiviruses in drugs and vaccines.

Climate Change and Emerging Infectious Diseases

Global warming is the rise in average surface temperature of the earth observed in the last decades (+0.6°C in the last half century; IPPC, 2007). This observation applies also specifically to Europe (EEA, 2012) and North America (Karoly et al., 2003), territories particularly concerned to WNV and BT virus (BTV) emergences, as it will be discussed later.

The increase has accelerated in recent years and is expected to go faster, so that in the twenty-first century the average surface temperature of the earth is expected to rise in a range between +1.6 and +6°C, depending on the different scenarios considered. Although still a contentious issue, most scientists now accept that human activity has contributed to the observed global warming , mostly through emissions of greenhouse gases produced by industrial activity and consumption of fossil fuels for transport, energy production, etc.

The impact of global warming on the environment is the subject of numerous recent studies, based on predictive modeling scenarios depending on the level of emissions. It is not the purpose of this article to review these models, which are described extensively elsewhere IPPC, 2001, 2007), but to highlight some future climate trends, mainly in Europe, the scenario for some recent unusual observations regarding arboviral diseases. For this, I will rely on data from the European project PRUDENCE (Prediction of Regional scenarios and Uncertainties for Defining EuropeaN Climate change risks and Effects; Christensen, 2005). In addition to an increase in mean annual air temperature, predicted to be between 1.4 and 4.5°C, depending on the areas (the highest rise is expected to occur in the Iberian Peninsula), the study suggests there will be more droughts, more wildfires, heat waves will be more frequent, and all this will be particularly noticeable in southern Europe. Winters will become milder, and this will occur more rapidly in northern latitudes. As a result, frost will decrease, and minimum temperatures will rise.

4

IMPORTANT DISEASES OF BUFFALOES AND THEIR CONTROL MEASURES

Infectious diseases are caused by biological agents like viruses, bacteria and fungi, which can be transmitted from animal to animal. These diseases cause economic losses to farmers by mainly decreasing the milk production in dairy animals. It can be controlled or prevented by timely and proper vaccination of farm animals.

Vaccination is the administration of a vaccine to boost the immune system for protection from a disease. A vaccine (or immunization) is a way to build animal's natural immunity to a disease before the animals gets infected. Vaccine contains a small amount of the disease germ (virus or bacteria) or parts of the germ in a weakened or killed state, or proteins or toxins from the organism. In stimulating the body's adaptive immunity, they help to prevent sickness from an infectious disease. Following are some important viral and bacterial diseases along with their preventive measures in farm animals which causes economic loss to farmers and required vaccination at regular time interval.

FOOT-AND-MOUTH DISEASE (FMD)

Foot and mouth disease is a severe, highly infectious and contagious viral disease of livestock that has a significant economic impact. The disease affects cattle, swine, sheep, goats and other cloven-hoofed ruminants. The virus causes high fever lasting two to six days, followed by blisters inside the mouth and on the feet that may rupture and cause lameness.

Transmission:

- Infected animals newly introduced into a herd (carrying virus in their saliva, milk, semen, etc)
- Contaminated pens/buildings or contaminated animal transport vehicles
- Contaminated materials such as hay, feed, water, milk
- Contaminated clothing, footwear, or equipment etc

Symptoms:

- High fever
- Vesicles and ulcers on tongue, dental pad, oral mucosa and interdigital space
- Profuse salivation, lameness

Prevention:

- Vaccination should be given at 4 months of age and it should be repeated at every 6 months
- Isolation of all affected animals after detection
- Burning of contaminated fodder and bedding material
- Animal shed must be cleaned with caustic soda or formalin
- Calves should not be allowed to suckle affected mothers

Treatment: There is no specific treatment, generally recovers within 8-10 days.

Symptomatic treatments:

- Parenteral injection of broad spectrum antibiotics to prevent secondary bacterial infection and analgesic
- Mouth wash with 1% $KMNO_4$ or 2 % Sodium carbonate solution
- Apply boroglycerine on mouth lesions
- Foot wash with 2 % copper sulphate or 2 – 4 % sodium carbonate solution

Post FMD complication:

- Abortion
- Mastitis
- Infertility
- Panting

There are two types of diseases mainly seen in animals.

(A) CONTAGIOUS

(B) INFECTIOUS.

A) CONTAGIOUS: As the name its elf suggests, contagious diseases are those which spread by contact/nearness of the animal.

(B) INFECTIOUS :These are diseases which are caused due to infective agent. (Virus or Bacteria or Fungus).All contagious diseases are infectious but all infectious deseases are not contagious.

Diseases can also be classified as VIRAL BACTERIAL, FUNGAL, RICKETTSIAL, OR PROTOZOON, VIRAL DESEASES.

Important ones : RINDERPEST, FOOT AND MOUTH, EPHIMORAL DISEASE, Blue tongue , Infectious bovine rhinotracheitis Malignant Catarrhal fever, Rabies

Rinderpest disease : causes diarrhoea, high fever and death. Effective vaccine control has controlled this disease in animals since 1998. There are no outbreaks recorded since them.

Foot & Mouth Disease : It is a viral disease. It has four main strains observed in India.(A,O,C,Asia I). In the symptoms animal goes lame. There is profuse salivation, animal goes lame. This disease has severe economic impact, on animal production (milk & Traction/Transport) This disease does not result in mortality (except few cases in calves). Disease can be prevented by undertaking regular vaccination. There are two types of vaccines. Aluminium hydroxide Gel and oil adjuvant vaccine. Both the vaccines are effective and useful.1st dose of vaccine is given at 2-3 month age. There after vaccine is given every 3 months as per the advice of the Veterinarian.

EPHIMERAL DISEASE

This viral disease is transmitted by insects. Animal shows high fever (104-105 F) and pain in muscles and joints. Animal is unable to walk, eat and ruminate. In India there is no vaccine against this disease. Symptomatic treatment such as non steroidal anti inflammatory drugs, soda alycylus,, potassium iodide is given/ venously. The course of the disease is for 4 to 7 days. Animal recovers of its own. Good nursing care is essential.

I.B.R Infectious Bovine Rhinotracheitis

It is a new emerging disease causing abortion, repeat breeding and infertility. It is estimate that nearly 30% of cross breads are affected. It shows various clinical symptoms. Respiratory disease (I B R) Genital disorders (IPV)infectious pustular vulvo-vaginitis in heifers/cows/buffaloes and infectious pustular balanoposthitis (IPB) in bulls. Infections of eye, nervous symptoms, etc. In India it was first recorded in 1976. Diagnosis is done by serological examination. Prevention is done by using oil adjuvant vaccine.

Malignant Catarrhal Fever (MCF) :All ages of animals are suseptible. There is Eye and Nose discharges, high fever, gastroenteritis,dyspnoea. Muzzle is red. There is central nervous system involvement.

The disease is suspected to be transmitted by insects. Recovered animals may carry virus for some months. In per acute cases animal dies within one to three days. In most common form there is drop in the milk. Acute gastro enteritis, swollen nasal mucous memberance high fever (104 to 105F),odema of eyelids, muzzle is red. There is foetid diarrhoea. There may be severe nervous dipression. Necrosis of skin over udder and teats may also be seen. Urine may be red because of free lesion .In bladder. Disease runs a course of 3 to 7 days. There is no effective treatment. Oxytetracycline intravenously, vit A, and good nursing can be carried out. Quarantine and slaughter are suggested for control.

RABIES: This viral disease is caused due to bite of an infected animal such as Dog, Cat, Fox, Bat or other domestic animal with a penetrating bitting wounds. Virus travels through nerves. Effective prevention is done by tissue culture vaccine in those areas where there is nearness of forest. There is no treatment.

Bacterial Diseases

Amongst important diseases following one need prevention and effective control. Anthrax, Black Quarters, Haemorrhagic Septicaemia,Tuberculosis and Johne'S Disease, Brucellosis,isolation of animal and strict hygenic measures are needed to control.

For remaining diseases antimicrobials can be tried.

ANTHRAX

This disease is seen on very less number of occasions now a days after the invention of antibiotics esp. Penicillin . It is caused due to Bacillus Anthracis. Disease causes sudden death in animals without apparent symptoms. The carcase gets bloated. In contact animals can be given higher doses of Penicillin. Disease can be controlled by using spared vaccines as per Veterinarians advice in endemic areas. Where there is no disease recorded in last three years, clean vaccination is not advised by many State Govt authorities.

Transmission: Infection usually spread through the ingestion of contaminated soil, fodder or compound feed. Anthrax spores in the soil are very resistant and can cause disease when ingested even years after an outbreak. The spores are brought to the surface by wet weather, or by deep tilling, and when ingested or inhaled by ruminants the disease reappears. Where an outbreak occurs, carcasses must be disposed of properly, the carcass should not be open (exposure to oxygen will allow the bacteria to form spores) and premises should be quarantined until all susceptible animals are vaccinated.

Vaccination in endemic areas is very important. Although vaccination will prevent outbreaks veterinary services sometimes fail to vaccinate when the disease has not appeared for several years. But because the spores survive for such lengthy periods, the risk is always present.

Symptoms:

- Sudden rise in body temperature (104-108 ÚF)
- Difficult breathing
- Blood in milk
- Abortions in female
- Bleeding from natural openings like anus, nostrils, vulva, sudden death
- Sudden death (often within 2 or 3 hours of being apparently normal) is by far the most common sign
- Very occasionally some animals may show trembling, a high temperature, difficulty breathing, collapse and convulsions before death. This usually occurs over a period of 24 hours
- After death blood may not clot, resulting in a small amount of bloody discharge from the nose, mouth and other openings

Diagnosis:

- On the basis of clinical signs.
- Rod-shaped bacteria surrounded by a capsule are visible in blood smears made from surface blood vessels.
- Post-mortem examinations should not be undertaken on suspected anthrax cases (including any cow that has died suddenly for no apparent reason) until a blood smear has proved negative);
- If a carcass is opened accidentally, the spleen is usually swollen and there is blood stained fluid in all body cavities.

Prevention:

- Vaccination with Anthrax spore vaccine @ 1ml S/C every year before monsoon season
- Identification and isolation of affected animals
- Deep burial of dead animals
- Never conduct post mortem of dead animal suspected to anthrax
- Disinfection of cattle shed by using 10 % caustic soda or 5% phenyl

Treatment:

- Treatment is effective in initial stage of the disease
- Injection penicillin @ 10000 units/kg body weight intramuscular for 5-6 days
- Anti-Anthrax serum @ 100-200 ml I/V if available
- Supportive injection of antipyretic, antihistaminic and fluid therapy

Due to the rapidity of the disease treatment is seldom possible, although high doses of penicillin have been effective in the later stages of some outbreaks.

BLACK QUARTERS

Disease is caused by claustridium charivoei, an anerobic organisms young healthy animals are affected. Infection is mainly through small wounds. There is swelling on the scapular muscles or on quarters, which on pressing gives crepitating sound. There is high fever and animal goes lame. Immediately use of antibiotics locally and parentrally is effective. Prevention can be done by carrying vaccination in endemic areas before mansoon.

It is an infectious bacterial disease most commonly caused by bacteria called Clostridium chauvoei. It is seen in livestock all over the world, usually affecting cattle, sheep, and goats. It has been seen occasionally in found bison and deer. The disease occurs suddenly which makes difficult in successful treatment and the efficacy of the commonly used vaccine is also hampered.

This disease mostly affects when the cattle are between the ages of six months and two years, although it can occur when they are as young as two months. Mostly, cattle that have good feed intake and are well-conditioned tend to be the most susceptible to blackleg.

Transmission: The spores of the organism can remain in the soil for years in an inactive state, and return to their infectious form when consumed by grazing livestock. Contaminated pasture is a major source of these organisms, which are also found naturally in the intestines of animals. Cases can occur over many years in areas where soil or manure is contaminated with the bacteria, and it is exceedingly difficult to remove the spores from the environment.

Symptoms:

- High fever
- Lameness in affected legs
- Crepitating, hot and painful swelling over the affected legs

Prevention:

- Vaccination should be given at 4 months of age and must be repeated every year
- Isolation of all affected animals after detection
- Deep burial or burning of dead animals
- Avoid grazing in affected area

Treatment:

- Parenteral injection of penicillin @ 10000 units/kg body weight for 5-6 days intramuscularly.
- Injection of BQ antiserum if available
- Parenteral injection of antipyretics and antihistaminic
- Fluid therapy – Infusion of 5% DNS I/V as supportive therapy

HAEMORRHAGIC SEPTICAEMIA

This Disease is caused by organisms known as Pasteurella multocida. The infective agent enters through respiratory tract by inhalation or through feed/ water. There is upper respiratory tract infection involving large bronchi and lunge. There is high fever ,watery discharge from nostrils. Animal is off feed, there is odema under the neck. There is respiratory distress. Animal dies if not treated in time. Mostly Buffalo species are affected. Higher antibiotics Tetracyclines,Sulpha Trimethoprim,or Enrofloxacin may be tried in early phases. No oral medication be done. Disease can be prevented by undertaking regular vaccination (GEL TYPE VACCINE) twice a day (Before mansoon and in october).

It is one of the most economically important infectious disease. The disease occurs mainly in buffaloes and cattle, but also has been reported in goats, generally occurs during rainy season, extreme environmental conditions, malnutrition and long distance transportation. A wide variety of clinical signs have been described for HS in buffaloes and cattle. The incubation periods (the time between exposure and observable disease) for buffalo calves 4–10 months of age varies according to the route of infection. The incubation period is 12–14 hours, approximately 30 hours and 46–80 hours for subcutaneous infection, oral infection and natural exposure, respectively.

Transmission:

- By direct contact with infected animals and infected clothes, utensils etc.
- Cattle and buffalo become infected when they ingest or inhale the causative organism, which probably originates in the nasopharynx of infected animals. In endemic areas, up to 5% of cattle and water buffalo may normally be carriers
- The worst epidemics occur during the rainy season, in animals in poor physical condition
- Stresses such as a poor food supply are thought to increase susceptibility to infection, and close herding and wet conditions seem to contribute to the spread of the disease
- multocida can survive for hours and possibly days in damp soil or water; viable organisms are not found in the soil or pastures after 2–3 weeks

Symptoms:

- High fever (106 – 107°F)
- Swelling of the head, neck, and brisket occurs in nearly all cases
- Swelling of throat region (sub mendibular edema)
- Difficult breathing
- Lacrimation
- Recumbancy and death within 10-72 hours.

Prevention:

- Vaccination every year before monsoon
- Isolation of all affected animals after detection
- Burning or burial of dead animals
- Proper disposal of contaminated feed and water
- Disinfection of cattle shed

- Avoid long distance transportation
- Avoid crowding, especially during wet conditions, will also reduce the incidence of disease

Treatment:

- Treatment is effective if given in early stage
- Parentral injection of antibiotics, antipyretics and antihistamininc
- Fluid therapy- Infusion of 5 % DNS I/V as supportive therapy

BRUCELLOSIS

This disease is caused by Brucella abortus. It is a disease of matured animals. It causes great economic loss by causing abortions in pregnant animals. In bulls it causes orchitis. The causative agent enters into body mainly through food/water and through wound even through conjunctival mucous memberance. Organisms locate in gravid uterus and causes inflammation of cotyledons resulting into abartions. It has zoonotic importance.There is no effective treatment. Diagnosis is done by serological examination. Prevention is done by undertaking calf hood vaccination (in females) males are not vaccinatec. Serological examination of all animals isolation of affected animals and calf hood vaccination is advised.

Brucellosis infection is caused by Brucella abortus, in worldwide which causes of bovine abortion that may infect wild species also. Brucellosis is an important zoonosis and all precautions must be taken against acquiring infection. Veterinarians are especially at risk when removing retained placenta and examining aborted cows and the products of abortion. The condition in humans was commonly acquired by drinking unpasteurized contaminated milk.

Transmission: As Brucellosis is an important zoonotic disease, the bacteria are transmitted from animals to humans by ingestion through infected food products, direct contact with an infected animal, or inhalation of aerosols. The disease has been known by various names, including Mediterranean fever, Malta fever, gastric remittent fever, and undulant fever. Humans are accidental hosts, but brucellosis continues to be a major public health concern worldwide and is the most common zoonotic infection. Brucella organisms, which are small aerobic intracellular coccobacilli, localize in the reproductive organs of host animals, causing abortions and sterility. They are shed in large numbers in the animal's urine, milk, placental fluid, and other fluids.

Clinical Symptoms: Abortion during the last trimester with retention of placental, followed by poor lactation.

Diagnosis:

- Based on clinical signs such as abortions, but confirmation is made through serological tests
- Prescribed laboratory tests to isolate and identify the bacteria

Prevention and control: Surveillance using serological tests, as well as tests like milk ring test, can be used for screening and play an important role in campaigns to eliminate the disease. Individual animal testing for disease control purposes is also being practiced. In endemic areas, vaccination is often used to reduce the incidence of infection. Several vaccines are available that use modified live viruses.

Human brucellosis is best prevented by controlling the infection in animals. Pasteurisation of milk from infected animals is an important way to reduce infection in humans.

TUBERCULOSIS

It is caused by Mycobacterium tuberculosis. Infeetion is mostly oral, either through feed or water. The disease has chronic course. Animal becomes weak day by day. In the udder form ,milk is positive for organisms. Diagnosis is by all allergic tests carried out on skim. Positive animals are segregated. There is no treatment. It has zoonotic improtance.

John's Disease

It is caused due to Mycobacterium para-tuber culosis. It has chronic couse. Infection is through oral route. Animal goes on Wasting. There is chronic diarrhoea. In pregnant animal disease appears to be suppressed, how ever on calving the symptoms start again. Diagnosis is done on the basis of allergic tests. There is no treatment. Test, and segregate positive animals.

Protoxoon Disease

These disease are carried by Ticks or bitling insects. The important disease are Theilariasis, Babesiosis, Anaplasmasis, Trypanasomiasisn.

Theilariasis

This is transmitted by Tciks. In India we have Theilaria annulata and Th.mutans. Th.annulata is pathogenic while Th.mutans is non pathogenic. As the infective stage invades red blood cells, there is weakness ,anaemia,and drop in production. The disease runs in Acute, sub-acuteand chronic form. There is initially high fever, local lymphriode swelling. Animal is dull. Diagnosis is done on blood smear examination. Treatment includes BUTALEX (BUPARVAQUONE) 1.5mg/kg body wt 1/musular. Tetracyclines higher doses are also tried. Supportive treatment such as B.Complex,Iron,Dextrose are given. Blood transfusion/PCV transfusion is also effective.

Babesiosis

This is tick bowrne disease. We have Babesia bovis, B.bigermina. Animal shows high fever (uptoloyF). There is coffee coloured urine. Animal may die if not treated. Some animals, because of destruction of red blood cells may show jaundice.

Diagnosis is done on the examination of blood smears. For the treatment BERENIL (DIMENAZEN) @ 7 mgs/kg sub cut is given. IMIDOCARB 9IMIDOCARB DIPROPIONATE) @1-2 mg per kg 1/muscular is also given.

Anaplasmosis

This is insect bourne disease. It has two types. ANAPLASMA CENTRALE AND ANAPLASMA MARGINALE. There is high fever (104-105F). There is red coloured urine. Animal may show jaundice. Animal becomes anaernic. Diagnosis is done on blood smear examination. In the treatment higher doses of antibiotics. TETRACYCLINES @ 10 mgs/kg.body wt 1/muscular is effective.

Trypanoso Miasis

This disease is transmitted by bitting insects from one to another animal. Infected animal shows high fever, there is reduction in production. There is odema over jawl on the brisket. In some animals there is central nervous system involvement. Animal moves in circle, it may dash against wall. Diagnosis is done on blood smear examination. Treatment can be based on one of the following. BERENIL (DIMENAZEN) @7 mg/kg s/c Antricide Prosalt 5 mg/kg wt s/c, Antricide sulphate @ 5 mg/kg s/c, SAMORIN (ISOMETAMEDIUM) 0.5 to 2 mg per kg s/c CYMELARSAN 0.3 to 0.6 mg/kg s/c control of insects by spraying ACARICIDES is important.

PARASITIC DISEASES

Dairy animals suffer from different types of worms. They are Round Worms, Tape Worms, Flukes and Coccidiosis. Young animals suffer more as compared to adult animals. General symptoms include weakness,drop in production and reduced immunity. Intensity of symptoms depends upon worm load and its type. Infection due to Liverfluke/Coccidiosis causes more damage when present. Diagnosis is done on the examination of stools. There are specific effective drugs for the treatment. In broad spectrum anthelmintics. ALBENDAZOL,OXFENDAZOL,INVERMECTIN are suggested. They are effective. They need repeatation after three weeks. In case of coccidiosis sulpha drugs can be given. There are ECTOPARASITIC DISEASES which infect skin. They are due to LICE/MITES. They cause skin lesions. Loss of hair, irritation, suratching are the symptoms. Diagnosis of type of mite can be carried out by examination of skin scrapings. Specific drugs are given for specific infestation.

METABOLIC DISEASES

Dairy animals, due to their high yielding nature are likely to suffer from METABOLIC DISEASES. These are Milk Fever, Ketosis, Post Parturient Haemoglobin Urea and Hypomagneshimia.

Milk Fever

There is no fever in this disease. This disease is seen mostly during the period of one week before calving to a week after calving. There is acute deficiency of circulating blood calcium. This results in depression, drop in milk yield, and recumbancy. Diagnosis is done on the basis of typical clinical symptoms and history. Treatment of calcium borogluconate 25%solution 200 to 400 ml Intravenously gives quick results. Injection of ivt.D3 is also useful. Prevention can be carried out by anionic mixtures before calving for calcium absorption. They are Ammonium chloride @150 gms/day for 10 days before calving.

Ketosis :

In high yielding animals deficiency of stored energy results in utilising body fact. While energy is available in this process, certain other products such as Ketone bodies (keto acids) are formed. The disease usually occurs during the peak yield phase. Animal passes little urine,it has sweet odur to its breath,it drops in production and becomes weak. Diagnosis is done by examination of Urine,for ketone bodies. Now a days paper strips are available,. In the treatment DEXTROSE 20-50% 500 to 300 ml i/venously along with GLUCO-CORTICOIDS 1/muscularly 3-5 ml are effective. Use of propylene glycol (300-500 gms) orally once/twice a day is effective. Proper feeding in advance pregnancy for storage of energy prevents disease incidance.

Post Parturient Haemoglobinurea

This disease occurs after calving due to acute deficiency of circulating phosphorus. There is lipid peroxidation of Red blood cell, resulting in its breaking and causing haemoglobinurea. Treatment consists of giving sodium acid phosphate 20% 300-500 ml 1/venously along with oral dosing @100 gms/day. Injection of BOTROPASE, PARAMIND BENZOIC ACID are also effective.

Hypomagnesemia

Acute deficiency of magnesium, results in high yielding animals due to excessive intake of potassium (green tender leaves feeding during mansoon) Animal becomes restless, it collapses, it has froth at mouth, heart beats are loud. Immediate treatment by using 10% Magnesium sulphate 100-200 ml I/venously will save the animal.

MASTITIS

It is the inflammation of the Udder. The causative agent may be Virus, Bacteria,Fungus or Algae. Non infective cause is deficiency of circulating cirtrate. Causative agent enters through teat sphincter (opening) mostly after milking (by hand or Machine) When sphincter muscles are yet to close tightly. (45 minutes to 1 ½ hour after milking). Once entered , it has good media to multiply inside the udder. In per acute form animal may die. In Acute case there is high fever, udder is swollen and painful. There is watery milk. In sub acute cases there is little swelting, no fever but milk is not normal. In chronic cases, there is little abnormality in udder how-ever milk quantity as well as quality is depressed.

The economic losses due to sub-clinical mastitis are more than clinical mastitis. In this form of disease there is no outward abnormality in the udder. Milk is normal but reduced. Per acute, acute,sub acute form of disease depends upon pathogenicity of organisms their number and defence mechanism status of the udder. As the causative agent enters through teat opening, minor cracks on teat or udder. It is therefore necessary to strictly follow practice of clean milk production. This includes washing of udder, drying (by cloth) of udder, before milking is necessary. Milking by full hand (not by following thumb) is also necessary. Milking when done by hand should be done quickly-completly and comfortably(comforts to cow).

After milking all the teats should be dipped in new, non toxic, non irritant disinfectant such as INSTACLEAN (Benzalkonium chloride +0.1%1-6dihexane) is useful it has long residual effect and works even in the presence of organic matter. If this practice is followed, incidance of sub-clinical mastitis can be controlled. In advanced pregnant animals, at the time of drying, all the teats should be infused with antibiotics. This will eliminate chronic mastitis.In the treatment following principles be observed.

Palpate the udder know its feel (hotness,hardness)

Record animals body temperature.

Remove milk from affected milk as frequently as possible.

If udder is hot and painful to touch,foment by ice/or cold water. Give antiinflammatory drugs (NSAID) intra muscular.

After removal of milk push in suitable antibiotic onitment as per the advice of veterinarian.

Always try to prevent, control mastitis than to treat the case.

After calving (from 15th day to 200 days of calving)examine milk from each teat by simple cow-side C.M.T.(California Mastitis Test) once in 15 days. This will detect sub clinical mastitis in time and effective control measures can be taken.

Mastitis is a man made disease, it needs proper hygenic measures to be adopted for udder care for control.

The above mentioned infectious diseases are serious problem in animals, so special care should be taken to milch animals and should be vaccinated periodically with the opinion of veterinarians. Vaccination in the last trimester in pregnant animals prevents the calves from infection until 4 months of age, so vaccination of pregnant animals must be done. So it is recommended that to prevent economic loss from the diseases, the farmers should follow the vaccination schedule at regular time interval.

5

TETANUS IN HORSES

Tetanus also called as lock jaw condition with worldwide distribution is characterized by hyperesthesia, tetany and convulsions. Tetanus is known to cause high fatality in horses by production of exotoxins affecting domestic animals, wildlife, and humans. Production of tetanospasmin and tetanolysin are responsible for the pathology of tetanus.

Tetanus, or lockjaw, is an often-fatal disease caused by a neurotoxin produced by the bacteria Clostridium tetani. C. tetani is present in the gastrointestinal tracts and feces of horses and other animals. C. tetani spores can be found in soil everywhere but are particularly abundant in areas where the soil is contaminated with manure.

The C. tetani bacterium produces the potent toxin "tetanus neurotoxin" (TeNT). TeNT causes spasmodic paralysis due to inhibition of neurotransmission (signalling) between the spinal cord and neurons. Horses, regardless of breed, age, or sex, are more susceptible to tetanus than any other domestic animal.

Tetanus is caused by Clostridium tetani, gram positive rod (0.3–0.6 ìm wide and 3–12 ìm long) spore forming anaerobic neurotoxic bacteria that is common soil inhabitant the spores of the tetanus can persist in soil for many years (Quinn et al 2011). C. tetani spores are round and terminal, giving typical shape as "tennis racket" (0.7 ìm × 0.5 ìm). The portal of entry is mainly through the deep puncture wound. In horses puncture wounds of the hooves are the common sites of entry and in neonatal equids the umbilicus remains as the main source of entry. Horses are reportedly as the most susceptible domestic species to the effects of tetanus toxin (Constable et al 2016).

The disease is characterized by generalized muscular rigidity and spasms, hyperesthesia, convulsions, respiratory arrest, and death. The incubation period varies between 3 days and 4 weeks with occasional cases as long as several months after the infection appears. Tetanus bacteria localizes at the site of the introduction and release the tetanospasmin and tetanolysin.

Tetanolysin aids in the local tissue necrosis while as tetanospasmin disseminates into the systemic circulation where it impedes the spontaneous and nerve impulse evoked release of neurotransmitter, which results in the disinhibtion of gamma motor neurons. The diagnosis of equine tetanus is based on typical clinical signs, presence of wounds or history of surgical procedures that may predispose to disease (Mackay 2014).

Septicemia or sepsis (blood infection) is a whole-body disease in which bacteria circulate in the bloodstream. It is most commonly found in very young foals less than 4 weeks of age. Bacteria can spread through the blood to various organs, such as the lungs, intestines, eyes, central nervous system, bones, and joints causing lameness and joint swelling. The condition implies an extensive, whole body infection from single or multiple sources. The most common bacteria identified in septic foals is Escherichia coli (E. coli). Others that may be involved include Klebsiella, Enterobacter, Actinobacillus, Salmonella, Pseudomonas, and Streptococcus species. Some of these bacteria contain a toxin in their cell walls called endotoxin. When endotoxin is released into the bloodstream, or when whole bacteria are present in the blood, it causes the release of other chemicals in the body called cytokines. These cytokines cause the signs of depression, poor appetite, and fever.

The major risk factor for septicemia in foals is the failure to receive an adequate amount of good quality colostrum (the first milk produced by mares, which contains high levels of antibodies) during the first few hours after birth. Other factors that increase the risk of disease include a dirty environment, prematurity at birth (a weak foal), poor health and condition of the dam, a difficult birth, and the presence of disease-causing agents in the environment that the mare has no antibodies against.

- Risk Factors for Septicemia in Foals
- Difficult birth
- Failure to receive an adequate quality and quantity of colostral antibodies
- Gestational age of the foal (prematurity)
- Poor health and condition of the dam

- Presence of new disease-causing agents in the environment against which the mare has no antibodies

UNSANITARY ENVIRONMENTAL CONDITIONS

A foal with septicemia may show evidence of single or multiple organ dysfunction. The earliest sign may be that the foal is slightly depressed or does not act as lively as other foals. This progresses to a foal that will not eat and lies down frequently. The mare's udder is often distended with milk, indicating that the foal is not nursing with normal frequency.

In the advanced stage of illness (septic shock), the body becomes overwhelmed. Such foals are often severely depressed and dehydrated, can only rise with assistance, and have a rapid heart rate. Low blood pressure, shown by such signs as cold extremities and a weak pulse, is evident. Foals may have elevated or reduced body temperature. Gums become either bright red or pale. In the final stages, foals become unconscious and go into cardiac and respiratory arrest.

Diagnosis is based in part on a thorough physical examination and medical history. Laboratory tests are required to assess organ function and identify the type of bacteria responsible for the infection.

TREATMENT AND OUTLOOK

Foals suspected of being septic are usually given broad-spectrum antibiotics (antibiotics that are effective against a wide range of bacteria) to kill the bacteria causing the infection. In cases of neonatal sepsis, veterinarians often recommend plasma transfusions to help the foal's immune system by supplying antibodies that were not received in colostrum. In addition, intravenous fluids may be given to help counter the effects of bacterial toxins and infection. Many septic foals have low blood sugar, so a sugar solution is often added to the fluid treatment.

Nutritional support is also important. If the foal is not nursing adequately, it should be fed mare's milk or a milk substitute at 15 to 25% of its body weight over each 24-hour period. Sometimes, the foal does not tolerate a large volume of milk and must be fed small amounts very frequently. It may be necessary to insert a feeding tube through the nose in foals with a decreased suckle reflex. Foals should never be fed milk by a syringe, because there is a risk of inhaling the milk into the lungs (aspiration).

Other treatments that may be required include flushing out infected joints with sterile fluids and providing nasal oxygen or ventilation for foals with septic pneumonia. Surgical removal of infected umbilical cord remnants may be needed. Eye infections and ulcers on the cornea of the eye are treated with antibiotic and anti-inflammatory eye ointments. Recovery from neonatal sepsis depends on the severity and form of the infection. Current survival rates are 55 to 75% in equine intensive care units of large veterinary hospitals. An average of 1 to 4 weeks of intensive care should be expected. Early recognition and intensive treatment of foal sepsis improves the outcome. If the foal survives the initial problems, it has the potential of becoming a healthy adult.

CLINICAL SIGNS OF TETANUS

The tetanus has different forms in horses acute and subacute and clinically as localized and generalized form. In the acute form, spastic paralysis rapidly spreads from the head (muscles of mastication, ears, third eyelid) to the respiratory muscles and then to the limbs. Generalized convulsions are accompanied by sweating. Death can occur in 1–2 d because of respiratory failure. In the subacute forms, signs develop in 1–3 wk. Some animals may recover.

Hyperesthesia and prolapse of the third eyelid are common early signs. Eating and swallowing are difficult because of paralysis of the mastication muscles. Localized tetanus is rare and may precede generalized tetanus. It causes painful spasm of muscles adjacent to the wound site. Generalized tetanus has a difficult course and involves a large number of symptoms. The main clinical signs include, erect ears, flare nostrils, sneering, pump handle respiration, excessive salivation, quidding, constipation, retention of urine and saw horse stance with a global feature comparable to that of a wooden horse.

RECENT STUDIES

The recent studies conducted by few authors in perspective to tetanus have focused on the epidemiology and outcome of tetanus cases in equines. Ribeiro et al 2018 studied retrospectively 70 cases of equine tetanus and recorded high mortality rate (72.9%) with forty (57.1%) horses had been presented history of wounds or surgical procedures related with tetanus, represented mainly by lesions in the hind limbs (42.5%), front limbs (15.0%), umbilical infections (7.1%), castration (4.3%), and face wounds (4.3%). Hyperesthesia, limb spasticity, cervical stiffness, tetanic spasms, and restriction of jaw movement were the main consistent clinical signs.

Kay and Knottenbelt (2007) studied prevelance of tetanus at hospital 56 equids with a clinical diagnosis of tetanus were hospitalised and treated, and 8 equids with tetanus were subjected to euthanasia without treatment. The total number of tetanus cases represented 0.07% of the total patient load and 2% of all hospitalised patients.

Van Galen G et al (2017) in a analysis of 176 tetanus cases in horses concluded that mortality rate was 68.2%. Mortality rates in adults and foals were 68.4% and 66.7%, respectively Variables associated with survival in adults included: standing, normal intestinal sounds and defecation, voluntarily drinking, eating soft or normal food, lower heart and respiratory rates, high base excess on admission, longer diagnosis time, treatment and hospitalization delay, and mild severity grade.

Variables associated with death included: anorexia, dysphagia, dyspnea, low blood potassium concentration on admission, moderate and severe disease grading, development of dysphagia, dyspnea, recumbency and seizures during hospitalization, treatment with glycerol guaiacolate, intravenous fluids, and intravenous glucose solutions. Variables associated with survival in foals included standing on admission, voluntarily eating soft food and drinking, older age, and longer hospitalization delay.

TREATMENT, MANAGEMENT AND CONTROL

The goal to the treatment of tetanus is based on five basic principles:

1. Elimination of the causative agent done by the administration of parental antibiotics arrest the growth and multiplication of tetani and thus also multiplication of the neurotoxin as well as suppression of secondary infections. At the same time, the wound should be cleaned of any foreign material such as devitalised and necrotic tissue with hydrogen peroxide and local application of pencillin may facililate absorption of the toxin.

The antibiotic of first option is procaine-penicillin G at a dose of 20—100,000 IU.kg administered intravenously, intramuscularly or subcutaneously every 6—12 hours. Other antibiotics, which may also be effective are: tetracycline at a dose of 22 mg.kg–1 orally every eight hours, clindamycin at a dose of 3—10 mg.kg–1 every eight to twelve hours intravenously, intramuscularly or orally, and amoxicillin-clavulanate at a dose of 12 mg.kg–1 orally or subcutaneously every twelve hour

2. Neutralize Residual toxin. As the signs of tetanus of appeared the administration of antitoxin has a little value However, One hundred units of antitoxin per 1 kg of body weight are recommended (maximum dose of 20 000 IU) given by a slow, over ten minutes intravenous administration. Repeated dose of antitoxin is not recommended because of the risk of anaphylactic and anaphylactoid-type reactions.

3. Control muscle Spasms Various drugs have been administered. Cholorpromazine (0.4-0.8mg/kg BW intravenously, 1.0mg/kg bw intramuscularly

three to four times a daily) and acetyl promazine (0.05mg/kg BW twice daily) until severe signs subside. A combination of diazepam (0.01-0.4mg/kg) and xylazine (0.5-1.0mg/kg, intravenously or intramuscularly).

4. Hydration can be maintained by intravenous or stomach tube.

5. Supportive treatment patient is hypersensitive to auditory or tactile stimulus, and therefore it should be kept in a quiet, dark room ears packed with cotton with well bedded quarters with non-slippery floors. Feeding and watering devices should be high enough to allow the horse to use them without lowering the head. Slings may be useful for horses having difficulty standing or rising.

Prevention and treatment of tetanus in horses

Tetanus in horses is typically fatal. Prevention of tetanus is essential and lies in vaccination. As all horses are susceptible to tetanus, the tetanus vaccine is considered the most basic equine vaccine and is indicated in the vaccination protocol for all horses. The AAEP recommends tetanus as a core vaccination.

Control

Tetanus can be avoided by proper disinfection of skin and instruments while carrying any kind of castration and vaccination. Immunization is recommended annually with tetanus toxoid and is usually recommended for all horses. Mares should be vaccinated during the last 6 weeks of pregnancy and the foals vaccinated at 5 to 8 weeks of age. In high-risk areas, foals may be given tetanus antitoxin immediately after birth and every 2 to 3 weeks until they are 3 months old, at which time they can be given toxoid.

6

SYMPTOMS OF BOTULISM DISEASES

Botulism (or botulism poisoning) is a rare but very serious illness that transmits through food, contact with contaminated soil, or through an open wound. Without early treatment, botulism can lead to paralysis, breathing difficulties, and death.

There are three main types of botulism:

Infant botulism

Foodborne botulism

Wound botulism

Botulism poisoning is due to a toxin produced by a type of bacteria called Clostridium botulinum. Although very common, these bacteria can only thrive in conditions where there's no oxygen. Certain food sources, such as home-canned foods, provide a potent breeding ground.

SYMPTOMS OF BOTULISM

Symptoms of botulism can appear from six hours to 10 days after the initial infection. On average, symptoms of infant and foodborne botulism appear between 12 and 36 hours after eating contaminated food.

Early signs of infant botulism include:

- Constipation
- Difficulty feeding
- Tiredness

- irritability
- drooling
- drooping eyelids
- weak cry
- loss of head control and floppy movements due to muscle weakness
- paralysis

Signs of foodborne or wound botulism include:

- difficulty swallowing or speaking
- facial weakness on both sides of the face
- blurred vision
- drooping eyelids
- trouble breathing
- nausea, vomiting, and abdominal cramps (only in foodborne botulism)
- paralysis

Foodborne botulism is a serious, potentially fatal disease. However, it is relatively rare. It is an intoxication usually caused by ingestion of potent neurotoxins, the botulinum toxins, formed in contaminated foods. Person to person transmission of botulism does not occur.

Spores produced by the bacteria Clostridium botulinum are heat-resistant and exist widely in the environment, and in the absence of oxygen they germinate, grow and then excrete toxins. There are 7 distinct forms of botulinum toxin, types A–G. Four of these (types A, B, E and rarely F) cause human botulism. Types C, D and E cause illness in other mammals, birds and fish.

Botulinum toxins are ingested through improperly processed food in which the bacteria or the spores survive, then grow and produce the toxins. Though mainly a foodborne intoxication, human botulism can also be caused by intestinal infection with C. botulinum in infants, wound infections, and by inhalation.

MECHANISM

The toxin is the protein botulinum toxin produced under anaerobic conditions (where there is no oxygen) by the bacterium Clostridium botulinum.

Clostridium botulinum is a large anaerobic Gram-positive bacillus that forms subterminal endospores.

There are eight serological varieties of the bacterium denoted by the letters A to H. The toxin from all of these acts in the same way and produces similar symptoms: the motor nerve endings are prevented from releasing acetylcholine, causing flaccid paralysis and symptoms of blurred vision, ptosis, nausea, vomiting, diarrhea or constipation, cramps, and respiratory difficulty.

Botulinum toxin is broken into eight neurotoxins (labeled as types A, B, C [C1, C2], D, E, F, and G), which are antigenically and serologically distinct but structurally similar. Human botulism is caused mainly by types A, B, E, and (rarely) F. Types C and D cause toxicity only in other animals.

In October 2013, scientists released news of the discovery of type H, the first new botulism neurotoxin found in forty years. However, further studies showed type H to be a chimeric toxin composed of parts of types F and A (FA).

Some types produce a characteristic putrefactive smell and digest meat (types A and some of B and F); these are said to be proteolytic; type E and some types of B, C, D and F are nonproteolytic and can go undetected because there is no strong odor associated with them.

When the bacteria are under stress, they develop spores, which are inert. Their natural habitats are in the soil, in the silt that comprises the bottom sediment of streams, lakes, and coastal waters and ocean, while some types are natural inhabitants of the intestinal tracts of mammals (e.g., horses, cattle, humans), and are present in their excreta. The spores can survive in their inert form for many years.

Toxin is produced by the bacteria when environmental conditions are favourable for the spores to replicate and grow, but the gene that encodes for the toxin protein is actually carried by a virus or phage that infects the bacteria. Unfortunately, little is known about the natural factors that control phage infection and replication within the bacteria.

The spores require warm temperatures, a protein source, an anaerobic environment, and moisture in order to become active and produce toxin. In the wild, decomposing vegetation and invertebrates combined with warm temperatures can provide ideal conditions for the botulism bacteria to activate and produce toxin that may affect feeding birds and other animals. Spores are not killed by boiling, but botulism is uncommon because special, rarely obtained conditions are necessary for botulinum toxin production from C. botulinum spores, including an anaerobic, low-salt, low-acid, low-sugar environment at ambient temperatures.

Botulinum inhibits the release within the nervous system of acetylcholine, a neurotransmitter, responsible for communication between motor neurons and muscle cells. All forms of botulism lead to paralysis that typically starts with the muscles of the face and then spreads towards the limbs. In severe forms, botulism leads to paralysis of the breathing muscles and causes respiratory failure. In light of this life-threatening complication, all suspected cases of botulism are treated as medical emergencies, and public health officials are usually involved to identify the source and take steps to prevent further cases from occurring.

Botulinum toxin A, C, and E cleave the SNAP-25, ultimately leading to paralysis.

CAUSE

Botulism is a disease caused by the botulinum toxin, which is produced by the bacterium Clostridium botulinum.

Clostridium botulinum spores are common in the soil, and also in the gut of healthy normal cattle and other animals in tropical environments (which includes most of Queensland), where they are not a problem. Spores are the dormant form of the organism. Only the actively growing or 'vegetative' Clostridium botulinum bacteria produce botulinum toxin and it is the toxin that produces the disease. Clostridium botulinum spores will only germinate and grow under conditions where oxygen is totally excluded.

Botulism outbreaks can occur in intensively fed beef and dairy cattle when:

The feed is contaminated with botulinum toxin that has been produced by the botulism bacteria growing in rotting animal or vegetable material in the stored feed. Contamination is in the form of the actual toxin actively growing botulism bacteria, eaten in feed contaminated with rotting animal or vegetable matter, continue growth in the rumen and gut of cattle, producing toxin. This source of disease is called Toxicoinfectious botulism. Sufficient toxin is produced for the animal to develop botulism. In this way, a relatively small source of the botulism organism, like a dead snake in contaminated feed, can infect large numbers of cattle fed a mixed ration chicken litter, used as fertiliser on pastures and not incorporated into the soil properly, has been accessed by cattle and consumed.

DISTRIBUTION

Botulism is commonly seen in the phosphorus deficient areas of northern Australia, where it is often associated with cattle eating bones and carrion to satisfy a craving for phosphorus and/or protein.

Reports of botulism have become more frequent in parts of Queensland that are not phosphorus deficient. Most of these outbreaks have been in intensively fed beef and dairy cattle. An increase in intensive feeding practices has occurred in the dairy and beef industries in Queensland, so the increasing risk of botulism outbreak has simply mirrored the increase in these feeding practices.

CLINICAL SIGNS

Botulinum toxin is often reported as being one of the most potent toxins known to humans, with only a small quantity required to produce disease.

The toxin binds strongly to nerve endings, preventing nerve impulses proceeding to muscles. This typically leads to the type of paralysis where animals go floppy or flaccid because they cannot move their muscles.

Clinical symptoms of botulism vary dramatically depending on the dose of toxin to which cattle are exposed and any pre-existing immunity that may be present. Signs will vary from sudden deaths (animals collapse and die in a couple of hours) to a slowly progressive paralysis where animals may take days to die.

In the latter case, the first signs are cattle off their feed and water. They then develop a wobbly gait (staggers) and eventually go down. During the staggers stage, some cattle become aggressive. Not all cattle that develop botulism symptoms will die. Some mildly affected cattle will recover. Generally speaking, once cattle go down, their likelihood of recovery is poor.

Cattle affected by botulism do not develop a fever. They show no response to treatment for other common causes of 'downer cow syndrome' such as three-day sickness or milk fever. Cattle may progress to the stage where they have difficulty breathing and typically lie on their brisket with their hind legs stretched out behind them. Tongue paralysis may or may not be a feature of the disease (cattle cannot pull their tongue back in when it is pulled out of their mouth).

At post-mortem, there are no obvious signs other than those associated with being down. Evidence of carrion and bones in the gut is uncommon in intensively fed animals, as most do not have access to these sources of botulinum toxin, unlike animals from phosphorus--deficient areas.

Diagnosis of botulism is based on signs suggestive of the disease and the absence of other possible causes. Isolation of Clostidium botulinum organisms from the intestinal contents of an animal fed a prepared ration is highly suggestive but not conclusive evidence of the disease.

Tests for botulinum toxin can be used to identify the toxin in gut contents, blood from affected cattle and samples of suspect feed and water. The success rate of testing for toxin is variable and, in many cases, botulism diagnosis is made on clinical signs and by excluding all other possible causes.

Another test looks for antibodies to botulinum toxin in cattle that survive. This test may support the diagnosis if the vaccination status of the animals is known prior to the outbreak and the animals have not subsequently been vaccinated.

IMPACTS

The risk of botulinum toxin entering milk supplies from dairy cows suffering from botulism is considered insignificant. Cattle affected by botulism generally cannot be milked. In addition, the toxin is extremely unlikely to pass from blood into the cow's milk.

However, immediately notify milk factories if you suspect potentially affected cattle have contributed to your bulk milk supply.

If surviving beef or dairy cattle are fit, healthy and normal, there is no known risk of botulinum toxin remaining in the carcass and subsequently poisoning people. Reports of botulism associated with other sources of toxin in prepared feed fed in feedlots or self-feeders in paddocks continue in Queensland, other states and overseas. Some of the worst botulism outbreaks have involved dairy cows being fed total mixed rations based on silage. In some cases, producers have lost two-thirds of their dairy herd over a 2-week period.

HOW IT IS SPREAD

The usual source of toxin in intensively fed cattle is feedstuff contaminated with rotting animal or vegetable material.

Previous sources of rotting material in stored feed have included: dead snakes and possums in grain augers snakes and other animals in hay and silage (snakes and other animals are killed by mowers or hide in windrows of hay and are then incorporated deep inside large round bales of hay or piles of silage) mice, when mouse plagues result in large numbers of mice dying in stored feedstuffs and grain augers, especially at the end of a plague water, when animals that die in dams, tanks or troughs are left to decompose.

Once incorporated deep in piles of silage or hay bales, or buried in large amounts of grain (in a silo or grain auger), where oxygen is not present, botulism bacteria can grow and produce toxin. The toxin produced can subsequently leach into the surrounding feed.

Once produced, the toxin is quite stable and may remain in contaminated feed or water for some time. The time it takes to break down will depend on environmental conditions.

RISK PERIOD

Improperly made silage that rots instead of fermenting is an ideal breeding ground for botulism if spores are present. Other possible sources are Brewer's grains and other wet by-product feeds, like citrus pulp and cannery waste that have been allowed to rot.

Monitoring and action

If you suspect botulism, contact a veterinary practitioner immediately for help with diagnosis and treatment. The disease can result in rapid, significant losses.

A private veterinary practitioner should examine the affected animals and conduct post-mortem examinations to rule out other possible diseases, such as a mass poisoning. Notify a government veterinarian or stock inspector to rule out exotic diseases if losses are significant.

If you suspect that botulism is a cause of mortalities in intensively fed animals, carefully examine the current ration and water supplies for possible sources of the toxin. If you find an obvious source, remove it and any feed or water it may have contaminated. Wherever possible, obtain any new ration components from a different source and replace the current ration immediately. Avoid any major changes in ration composition to avoid digestive upsets. It is also essential to start a vaccination program immediately.

A vaccination program will be necessary to prevent ongoing losses after an outbreak in pastured cattle. Also pay attention to phosphorous and protein nutrition. This may assist in the longer term by addressing nutritional deficiencies that predispose cattle to the bone chewing that exposes them to the botulinum toxin.

CONTROL

Prevention

In high-risk situations, vaccination against botulism is the only effective way to prevent botulism from occurring.

It is illegal to feed poultry litter or any other animal matter to cattle in Australia. Animals may graze pasture that has been fertilised with chicken faeces or litter if it is ploughed into the soil or given time to be incorporated into the soil before the animals are put to pasture.

Vaccination and antibiotics

All cattle in the herd should be immediately vaccinated with a bivalent botulinum vaccine; traditional formalised 'two shot' vaccine is the cheapest. The choice of which vaccine to use may depend on what is available at the time in sufficient quantities to vaccinate all your cattle.

Where cattle have been vaccinated early in the course of the outbreak, re-vaccination after 2 weeks has generally led to a cessation of new cases by significantly boosting the immune response. This recommendation for 2 weeks between first and second vaccinations differs from label recommendations on most botulinum vaccines. It is done in an attempt to accelerate the immune response to botulism and prevent further cases occurring. Field experience suggests it is an effective way of halting botulism outbreaks. If the revaccination interval is reduced, you should consult with the product manufacturer on whether they recommend a third shot of vaccine for long-term protection. This will depend on the type of vaccine used.

The use of oral antibiotics to prevent toxicoinfectious botulism from occurring has never been scientifically validated. The use of sufficient oral doses of antibiotics to kill Clostridium botulinum organisms may lead to significant digestive upsets in treated cattle.

Quality feedstuffs

Ensuring that feedstuffs are not contaminated with botulinum toxin is part of the assurance systems now implemented in all commercial feedlots and most dairies.

Vermin control during the preparation and storage of animal feedstuffs is very important.

Addressing nutritional deficiencies

In extensively grazed areas, the correction of any dietary deficiencies will help prevent bone chewing by cattle. However, it will not be a sufficient strategy on its own, so it should always be accompanied by an appropriate vaccination program. Addressing nutritional deficiencies will often result in some production improvements as well.

Avoiding problems with chicken litter on pastures

Chicken litter has been demonstrated many times to be a potent source of botulinum toxin. Where chicken litter is used as a fertiliser on pastures, it should be used only in situations where it can be incorporated into the soil immediately after being spread such as when pastures are being renovated. Cattle should not be allowed access to piles of stored chicken litter because they will eat it.

There is a chicken litter feeding ban under Queensland law; producers who knowingly feed their animals chicken litter or fail to take every reasonable measure to deny their animals access to the litter, may be prosecuted.

Vaccination the only long-term prevention strategy

Beef and dairy producers in Queensland who feed their cattle a prepared ration, especially those based on silage or by-products such as brewer's grains, should vaccinate their cattle against botulism.

Large areas of Queensland are phosphorous deficient and botulism is a common disease of grazing livestock in these areas. In addition, some areas are seasonally phosphorus deficient, such as in wet seasons in coastal north Queensland where lush growth results in a relative phosphorus deficiency in pastures. Vaccination is the most reliable means of preventing botulism losses in these areas.

Whenever possible, cattle should be vaccinated well before any suspected period of risk. Vaccinating well before the ration is introduced is recommended.

Various effective botulism vaccines are available. Some newer vaccines require only a single shot (rather than the traditional 2 shots a month apart). Both the 1-shot and 2-shot vaccines produce a similar end result, and the decision on which type of vaccine to use depends largely on product cost and convenience.

All vaccines require boosters to maintain protective levels of immunity. Consult package information or the vaccine manufacturer for advice on the timing of booster vaccinations.

Symptoms of foodborne botulism

Botulinum toxins are neurotoxic and therefore affect the nervous system. Foodborne botulism is characterized by descending, flaccid paralysis that can cause respiratory failure. Early symptoms include marked fatigue, weakness and vertigo, usually followed by blurred vision, dry mouth and difficulty in swallowing and speaking. Vomiting, diarrhoea, constipation and abdominal swelling may also occur. The disease can progress to weakness in the neck and arms, after which the respiratory muscles and muscles of the lower body are affected. There is no fever and no loss of consciousness.

The symptoms are not caused by the bacterium itself, but by the toxin produced by the bacterium. Symptoms usually appear within 12 to 36 hours (within a minimum and maximum range of 4 hours to 8 days) after exposure. Incidence of botulism is low, but the mortality rate is high if prompt diagnosis and appropriate, immediate treatment (early administration of antitoxin and intensive respiratory care) are not given. The disease can be fatal in 5 to 10% of cases.

EXPOSURE AND TRANSMISSION

Foodborne botulism

C. botulinum is an anaerobic bacterium, meaning it can only grow in the absence of oxygen. Foodborne botulism occurs when C. botulinum grows and produces toxins in food prior to consumption. C. botulinum produces spores and they exist widely in the environment including soil, river and sea water.

The growth of the bacteria and the formation of toxin occur in products with low oxygen content and certain combinations of storage temperature and preservative parameters. This happens most often in lightly preserved foods and in inadequately processed, home-canned or home-bottled foods.

C. botulinum will not grow in acidic conditions (pH less than 4.6), and therefore the toxin will not be formed in acidic foods (however, a low pH will not degrade any pre-formed toxin). Combinations of low storage temperature and salt contents and/or pH are also used to prevent the growth of the bacteria or the formation of the toxin.

The botulinum toxin has been found in a variety of foods, including low-acid preserved vegetables, such as green beans, spinach, mushrooms, and beets; fish, including canned tuna, fermented, salted and smoked fish; and meat products, such as ham and sausage. The food implicated differs between countries and reflects local eating habits and food preservation procedures. Occasionally, commercially prepared foods are involved.

Though spores of C. botulinum are heat-resistant, the toxin produced by bacteria growing out of the spores under anaerobic conditions is destroyed by boiling (for example, at internal temperature greater than 85 °C for 5 minutes or longer). Therefore, ready-to-eat foods in low oxygen-packaging are more frequently involved in cases of foodborne botulism.

Food samples associated with suspect cases must be obtained immediately, stored in properly sealed containers, and sent to laboratories in order to identify the cause and to prevent further cases.

Infant botulism

Infant botulism occurs mostly in infants under 6 months of age. Different from foodborne botulism caused by ingestion of pre-formed toxins in food, it occurs when infants ingest C. botulinum spores, which germinate into bacteria that colonize in the gut and release toxins. In most adults and children older than about 6 months, this would not happen because natural defences in intestines that develop over time prevent germination and growth of the bacterium. C. botulinum in infants include constipation, loss of appetite, weakness, an altered cry and a striking loss of head control. Although there are several possible sources of infection for infant botulism, spore-contaminated honey has been associated with a number of cases. Parents and caregivers are therefore warned not to feed honey to the infants before the age of 1 year.

Wound botulism

Wound botulism is rare and occurs when the spores get into an open wound and are able to reproduce in an anaerobic environment. The symptoms are similar to the foodborne botulism, but may take up to 2 weeks to appear. This form of the disease has been associated with substance abuse, particularly when injecting black tar heroin.

Inhalation botulism

Inhalation botulism is rare and does not occur naturally, for example it is associated with accidental or intentional events (such as bioterrorism) which result in release of the toxins in aerosols. Inhalation botulism exhibits a similar clinical footprint to foodborne botulism. The median lethal dose for humans has been estimated at 2 nanograms of botulinum toxin per kilogram of bodyweight, which is approximately 3 times greater than in foodborne cases.

Following inhalation of the toxin, symptoms become visible between 1–3 days, with longer onset times for lower levels of intoxication. Symptoms proceed in a similar manner to ingestion of botulinum toxin and culminate in muscular paralysis and respiratory failure.

If exposure to the toxin via aerosol inhalation is suspected, additional exposure to the patient and others must be prevented. The patient's clothing must be removed and stored in plastic bags until it can be washed thoroughly with soap and water. The patient should shower and be decontaminated immediately.

Other types of intoxication

Waterborne botulism could theoretically result from the ingestion of the pre-formed toxin. However, as common water treatment processes (such as boiling, disinfection with 0.1% hypochlorite bleach solution) destroy the toxin, the risk is considered low.

Botulism of undetermined origin usually involves adult cases where no food or wound source can be identified. These cases are comparable to infant botulism and may occur when the normal gut flora has been altered as a result of surgical procedures or antibiotic therapy. Adverse effects of the pure toxin have been reported as a result of its medical and/or cosmetic use in patients, see more on 'Botox' below.

'Botox'

The bacterium C. botulinum is the same bacterium that is used to produce Botox, a pharmaceutical product predominantly injected for clinical and cosmetic use. Botox treatments employ the purified and heavily diluted botulinum neurotoxin type A. Treatment is administered in the medical setting, tailored according to the needs of the patient and is usually well tolerated although occasional side effects are observed.

DIAGNOSIS AND TREATMENT

Diagnosis is usually based on clinical history and clinical examination followed by laboratory confirmation including demonstrating the presence of botulinum toxin in serum, stool or food, or a culture of C. botulinum from stool, wound or food. Misdiagnosis of botulism sometimes occurs as it is often confused with stroke, Guillain-Barré syndrome, or myasthenia gravis.

Antitoxin should be administered as soon as possible after a clinical diagnosis. Early administration is effective in reducing mortality rates. Severe botulism cases require supportive treatment, especially mechanical ventilation, which may be required for weeks or even months. Antibiotics are not required (except in the case of wound botulism). A vaccine against botulism exists but it is rarely used as its effectiveness has not been fully evaluated and it has demonstrated negative side effects.

Prevention

Prevention of foodborne botulism is based on good practice in food preparation particularly during heating/sterilization and hygiene. Foodborne botulism may be prevented by the inactivation of the bacterium and its spores in heat-sterilized (for example, retorted) or canned products or by inhibiting bacterial growth and toxin production in other products. The vegetative forms of bacteria can be destroyed by boiling but the spores can remain viable after boiling even for several hours. However, the spores can be killed by very high temperature treatments such as commercial canning.

Commercial heat pasteurization (including vacuum packed pasteurized products and hot smoked products) may not be sufficient to kill all spores and therefore the safety of these products must be based on preventing bacterial growth and toxin production. Refrigeration temperatures combined with salt content and/ or acidic conditions will prevent the growth of the bacteria and formation of toxin.

The WHO Five Keys to Safer Food serve as the basis for educational programmes to train food handlers and educate the consumers. They are especially important in preventing food poisoning.

The Five Keys are:

- keep clean
- separate raw and cooked
- cook thoroughly
- keep food at safe temperatures
- use safe water and raw materials.

WHO'S RESPONSE

Botulism outbreaks are rare but are public health emergencies that require rapid recognition to identify the disease source, distinguish outbreak types (between natural, accidental or potentially deliberate), prevent additional cases and effectively administer treatment to affected patients.

Successful treatment depends significantly on early diagnosis and the rapid administration of the botulinum antitoxin.

WHO's role in responding to outbreaks of botulism that may be of international concern is as follows.

Surveillance and detection: WHO supports the strengthening of national surveillance and international alert systems to ensure rapid local outbreak detection and an efficient international response. WHO's main tool for these activities of surveillance, coordination and response is the use of the International Network of Food Safety Authorities (INFOSAN) which links national authorities in Member States in charge of managing food safety events. This network is managed jointly by FAO and WHO.

Risk assessment: WHO response is based on a risk assessment methodology that includes consideration of whether the outbreak is natural, accidental, or, possibly, intentional. WHO also provides scientific assessments as basis for international food safety standards, guidelines and recommendations developed by the Codex Alimentarius Commission.

Containment at the disease source: WHO coordinates with national and local authorities in order to contain outbreaks at their source.

Delivery of assistance: WHO coordinates between international agencies, experts, national laboratories, airlines and commercial organizations to mobilize response equipment, materials and supplies, including the provision and administration of botulinum antitoxin.

Definition

Botulism is an acute, progressive condition caused by botulinum toxin, a natural poison produced by the spore-forming bacteria Clostridium botulinum. Exposure to the botulinum toxin usually occurs from eating contaminated food although, in infants, it may be caused by specific types of clostridia obtained from soil or inhaled spores, causing growth of the bacteria in the infant's intestine. Botulinum toxin is a neurotoxin that blocks the ability of motor nerves to release acetylcholine, the neurotransmitter that relays nerve signals to muscles, a process that may result in unresponsive muscles, a condition known as flaccid paralysis. Breathing may be severely compromised in progressive botulism because of failure of the muscles that control the airway and breathing.

Botulism occurs only rarely, but its high fatality rate makes it a great concern for those in the general public and in the medical community. Clinical descriptions of botulism reach as far back in history as ancient Rome and Greece. However, the relationship between contaminated food and botulism was not defined until the late 1700s. In 1793 the German physician, Justinius Kerner (1786–1862), deduced that a substance in spoiled sausages, which he called wurstgift (German for sausage poison), caused botulism. The toxin's origin and identity remained vague until Emile van Ermengem (1851–1932), a Belgian professor, isolated Clostridium botulinum in 1895 and identified it as the source of food poisoning.

Three types of botulism have been identified: food-borne, wound, and infant botulism. The main difference between types hinges on the route of exposure to the toxin. Food-borne botulism accounts for 25 percent of all botulism cases and can usually be traced to eating contaminated home-preserved food. Infant botulism accounts for 72 percent of all cases. About 98 percent of infants recover with proper treatment. Although domestic food poisoning is a problem worldwide, concern is growing regarding the use of botulism toxin in biological warfare. At the end of the twentieth century 17 countries were known to be developing biological weapons, including the culture of botulism toxins.

Transmission

Botulism is not spread from one individual to another, but through exposure to the deadly botulinum toxin, a natural poison produced by certain Clostridium bacteria that may be found in preserved, especially canned, foods and sometimes in the intestines of infants. Botulism spores can cause widespread illness if introduced into the environment.

Demographics

Botulism occurs worldwide, with 90 percent of the comparatively rare cases occurring in the United States. Approximately 110 cases of botulism are reported annually in the United States, with 50 percent of cases in California alone. Infant botulism accounts for 72 percent of all cases, far exceeding both food-borne and wound botulism. Food-borne botulism accounts for 25 percent of all cases, primarily due to eating contaminated home-preserved food.

How Can I Prevent Botulism?

- In most cases, botulism is easy to prevent. You can reduce your risk with the following preventative measures:
- Follow proper techniques when canning food at home, ensuring you reach adequate heat and acidic levels.
- Be cautious of any fermented fish or other aquatic game foods.
- Throw away any open or bulging cans of commercially prepared food.
- Refrigerate oils infused with garlic or herbs.
- Potatoes cooked and wrapped in aluminum foil can create an oxygen-free environment where botulism can thrive. Keep these hot or refrigerate immediately.
- Boiling foods for 10 minutes will destroy botulism toxin.

Causes and symptoms

Toxins produced by the bacterium Clostridium botulinum are the main culprit in botulism. Other members of the Clostridium genus can produce botulinum toxin, namely C. argentinense, C. butyricum, and C. baratii, but these are minor sources. To grow, these bacteria require a low-acid, oxygen-free environment that is warm (40–120°F or 4.4–48.8°C) and moist. Lacking these conditions, the bacteria transform themselves into spores that, like plant seeds, can remain dormant for years. Clostridia and their spores exist all over the world, especially in soil and aquatic sediments. They do not threaten human or animal health until the spores encounter an environment that favors growth. The spores then germinate, and the growing bacteria produce the deadly botulism toxin.

Scientists have discovered that clostridia can produce at least seven types of botulism toxin, identified as A, B, C, D, E, F, and G. Humans are usually affected by A, B, E, and very rarely F; infants are affected by types A and B. Domesticated animals such as dogs, cattle, and mink are affected by botulism C toxin, which also affects birds and has caused massive die-offs in domestic bird flocks and wild waterfowl. Botulism D toxin can cause illness in cattle, and horses succumb to botulism A, B, and C toxin. There have been no confirmed cases of human or animal botulism linked to the G toxin.

In humans, botulinum toxin latches onto specific proteins in nerve endings and irreversibly destroys them. These proteins control the release of acetylcholine, a neurotransmitter that stimulates muscle cells. With acetylcholine release blocked, nerves are not able to stimulate muscles. Ironically, this action of the botulinum toxin has given it a beneficial niche in the world of medicine. Certain medical disorders are characterized by involuntary and uncontrollable muscle contractions. Medical researchers have discovered that injecting a strictly controlled dose of botulinum toxin into affected muscles inhibits excessive muscle contractions. The muscle is partially paralyzed and normal movement is retained.

Human botulism (caused by botulism toxins A, B, and E) may stem from contaminated food, wound contamination, or the intestinal botulism toxin found in infants. Each produces multiple symptoms as follows:

Food-borne botulism. Food that has been improperly preserved or stored can harbor botulinum toxin-producing clostridia. Canned or jarred baby food has also been known to cause botulism. Symptoms of food-borne botulism typically appear within 18 to 36 hours of eating contaminated food, with extremes of four hours to eight days. Initial symptoms include blurred or double vision and difficulty swallowing and speaking. Possible gastrointestinal problems include constipation , nausea , and vomiting . As botulism progresses, the victim experiences weakness or paralysis, starting with the head muscles and progressing down the body. Breathing becomes increasingly difficult. Without medical care, respiratory failure and death are very likely.

Infant botulism. Infant botulism was first described in 1976. Unlike adults, infants younger than 12 months are vulnerable to C. botulinum colonizing the intestine. Infants ingest spores in honey or simply by swallowing spore-containing dust or dirt.

The spores germinate in the large intestine and, once colonized, toxin is produced and absorbed into the infant's body from the entire intestinal tract. The first symptoms include constipation, lethargy, and poor feeding. As infant botulism progresses, sucking and swallowing (thus eating) become difficult. A nursing mother will often notice her own breast engorgement as the first sign of her infant's illness. The baby suffers overall weakness and cannot control head movements. Because of the flaccid paralysis of the muscles, the baby appears floppy. Breathing is impaired, and death from respiratory failure is a very real danger.

Wound botulism. Confirmed cases of wound botulism have been linked to trauma such as severe crush injuries to the extremities, surgery, and illegal drug use. Wound botulism occurs when Clostridia colonize an infected wound and produce botulinum toxin. The symptoms usually appear four to 18 days after an injury occurs and are similar to food-borne botulism, although gastrointestinal symptoms may be absent.

Stranded Cladophora mat accumulations are common along shorelines of the Great Lakes during warm summer months, directly affecting visitor enjoyment due to noxious odors from decomposing mats and potentially causing economic losses from reduced recreational days. Numerous studies have shown that Cladophora in nearshore waters can affect beach water quality since the algal mats are known to harbor high levels of the enteric bacteria: E. coli and enterococci, with counts often exceeding 100,000 CFU/g. Aside from these commensal bacteria, various human pathogens, such as shiga toxin-producing E. coli, Salmonella, Shigella, and Campylobacter have also been recovered in Cladophora mat. It has been postulated that high levels of these bacteria are due to in situ growth as the mats are a natural substrate for a variety of micro- and macroorganisms.

There is growing speculation that Cladophora may have a role in bird botulism outbreaks a fatal bacterial disease that threatens lake and wetland birds. This disease is caused by ingestion of neurotoxins produced by Clostridium botulinum bacteria. The neurotoxin typeE is common in fish-eating lake birds, causing major die-offs in the lower Great Lakes. Some of the commonly affected birds include ring-billed gulls, herring gulls, double-crested cormorants, horned grebes, red-necked grebes, mergansers, common loons, and scoters. In Lake Michigan, botulism outbreaks have been more common in recent years, with beaches around Sleeping Bear Dunes National Lakeshore (SLBE) identified as the epicenter.

C. botulinum bacterium is nearly ubiquitous in the environment â€" especially, in river, wetland, and lake sediments; however, the mechanism by which the disease manifests in birds remains poorly understood. Fish (e.g., gobies) and macroinvertebrates (e.g., Quagga mussels and other benthic organisms), both identified as part of the food web, have been suggested to play a critical role in disease transmission. High bird mortality at SLBE coincides with massive Cladophora accumulations in benthic habitats at a time when the algae shifts to a heterotrophic condition. Thus, it has been postulated that decaying Cladophora during late summer likely create nutrient-rich, anaerobic conditions that are conducive for the sustenance of C. botulinum. One of the pathways might include vegetative cells/spores and/or the toxin itself via the various intermediaries (sediment-Cladophora-invertebrates-fishes) ultimately to birds for piscivorous species. GLSC scientists and their collaborators are studying the role of Cladophora in supporting C. botulinumgrowth; mussels and other macroinvertebrates in botulism disease transmission; and whether C. botulinum associated with Cladophora mats produces the potent typeE neurotoxin. Generally, current monitoring programs for beach health and botulism outbreaks are relatively less frequently. Therefore, a more frequent/robust monitoring would provide valuable data that could be used to develop better methods to predict and prevent beach closures and botulism outbreaks.

What Is Botulism?

There are seven types of botulinum toxin, each designated by a letter from A through G. Only four types (A, B, E, and F) make people sick. There are three forms of naturally occurring human botulism: infant botulism, food-borne botulism, and wound botulism. Inhalation (in-huh-LAY-shun) botulism is an additional form of the illness that could possibly be spread through the air intentionally by man. Clostridium botulinum is commonly found in soil and grows best in low-oxygen environments. The bacteria form spores that remain dormant, or inactive, waiting for conditions that allow them to grow. The spores can exist everywhere, and people might eat food containing these spores without becoming sick. When conditions are right, however, the spores can activate and produce toxin.

Food contaminated with the toxin is the culprit in cases of food-borne botulism; most outbreaks stem from food that is improperly cooked or incorrectly canned or preserved. In food-borne botulism the toxin itself is swallowed, but in cases of infant botulism an infant swallows the spores and they then activate in the intestine and produce toxin. It is believed that the infant intestine lacks enough protective intestinal bacteria, stomach acid, and immune globulins to prevent the spores from activating. Wound botulism is rare and develops when bacteria infect a wound and grow, producing the toxin. Bioterrorists have attempted to aerosolize botulinum toxin without success, but the threat of inhalation botulism as a biological weapon has raised concerns.

How Common Is Botulism?

More than 100 cases of botulism are reported in the United States each year. Infant botulism accounts for about 72 percent of reported cases and food-borne botulism for about 25 percent. Wound botulism is the rarest form, but health officials have noted an increase in this type in California, a rise they attribute to the intravenous use of illegal drugs from Mexico.

- Toxins are poisons that harm the body.
- Spores are a temporarily inactive form of a germ enclosed in a protective shell.
- Immune globulins (ih-MYOON GLAH-byoo-linz), also called gamma globulins, are the proteins of which antibodies are composed.
- Aerosolize (AIR-o-suh-lize) is to put something, such as a medication, in the form of small particles or droplets that can be sprayed or released into the air.
- Intravenous (in-tra-VEE-nus) means within or through a vein. For example, medications, fluid, or other substances can be given through a needle or soft tube inserted through the skin's surface directly into a vein.

What Are the Signs and Symptoms of Botulism?

The classic symptoms of botulism include blurred or double vision , droopy eyelids, slurred speech, difficulty in swallowing, dry mouth, and muscle weakness. These symptoms appear when the toxin interrupts nerve impulses to the muscles, which paralyzes the muscles. If untreated, paralysis may progress to involve the arms, legs, trunk, and muscles of the respiratory tract. In food-borne botulism, symptoms generally begin 18 to 36 hours after eating the contaminated food, but they can occur after as little as 6 hours or as much as 10 days. Infants with botulism may appear drowsy or sluggish, not eat well, be constipated, and have a weak cry and muscle weakness. In infants, it can take 3 to 30 days for the symptoms to appear and progress.

How Do Doctors Make the Diagnosis?

Laboratory tests can detect the toxin in blood, stool (bowel movements), or wound samples. Because the symptoms of botulism are similar to those of stroke and several other nerve diseases, doctors also may order a brain scan, spinal tap , or other nerve- and muscle-function tests to check for other possible causes. The doctor may ask whether the patient has eaten any home-canned foods and, if so, order tests on the suspect food.

How Is Botulism Treated?

For foodborne and wound botulism, a doctor administers an antitoxin as soon as possible after diagnosis. In infants, a treatment known as botulism immune globulin blocks the actions of neurotoxins circulating in the blood.

Severe cases of botulism may require the use of a ventilator to help support breathing. Recovery may take weeks or months. Long-term therapy and rehabilitation may also be necessary in severe cases. There's a vaccine for botulism, but it's not common, as its effectiveness hasn't been fully tested and there are side effects.

The U.S. Centers for Disease Control and Prevention (CDC) has a supply of antitoxin against botulism. The antitoxin can slow or halt the damage caused by botulinum toxin. The sooner it is given, the more effective it is in easing the symptoms. To help rid the body of the toxin, doctors sometimes cause vomiting or use enemas . In severe cases, patients who are unable to breathe well enough on their own might need a ventilator (VEN-tuh-lay-ter), a machine that can help a person breathe for several weeks. Wound botulism might require surgery to remove the source of the toxin-producing bacteria.

What Are the Complications and Duration of the Disease?

Most people with botulism require hospitalization, but they typically recover after weeks or perhaps months of care. Paralysis of the respiratory muscles can lead to pneumonia. Even after recovery, some patients may be tired and feel short of breath.

- double vision is a vision problem in which a person sees two images of a single object.
- respiratory tract includes the nose, mouth, throat, and lungs. It is the
- pathway through which air and gases are transported down into the lungs and back out of the body.
- stroke is a brain-damaging event usually caused by interference with blood flow to the brain. A stroke may occur when a blood vessel supplying the brain becomes clogged or bursts, depriving brain tissue of oxygen. As a result, nerve cells in the affected area of the brain, and the specific body parts they control, do not properly function.
- spinal tap, also called a lumbar puncture, is a medical procedure in which a needle is used to withdraw a sample of the fluid surrounding the spinal cord and brain. The fluid is then tested, usually to detect signs of infection, such as meningitis, or other diseases.

- antitoxin (an-tih-TOK-sin) counteracts the effects of toxins, or poisons, on the body. It is produced to act against specific toxins, like those made by the bacteria that cause botulism or diphtheria.
- enemas (EH-nuh-muhz) are procedures in which liquid is injected through the anus into the intestine, usually to flush out the intestines.
- pneumonia (nu-MO-nyah) is inflammation of the lung, usually caused by infection.

How Can Botulism Be Prevented?

Although botulinum toxin is extremely potent, it can be destroyed easily. Heating food and drinks to an internal temperature of 185 degrees

Fahrenheit for at least 5 minutes will detoxify them. Boiling home-canned foods before eating them also lessens risk. Health officials advise using a pressure cooker and high temperatures—about 250 degrees Fahrenheit—to kill the spores when canning or preserving foods at home. It also is best to avoid eating commercially prepared foods from cans that are swollen, punctured, or leaking. Because honey can contain spores of Clostridium botulinum, doctors advise that infants younger than 1 year not be given this sweetener. Breast-feeding can help protect against infant botulism. Receiving prompt medical care for infected wounds and not injecting street drugs can help prevent wound botulism.

For Good and Evil

Botulinum toxin is the first biological toxin licensed for treatment of human diseases. The U.S. government approves its use to relax painfully cramped or tight muscles and as an ingredient in medications for the treatment of migraine headache, chronic lower back pain, stroke, brain injury, and cerebral palsy. Botulinum toxin injections (called Botox) can paralyze muscles that cause the skin to wrinkle. Their use is popular among people looking to maintain a youthful appearance. Ironically, the same substance has the potential to cause mass destruction if it is dispersed into the air or introduced into the food supply. The U.S. government developed botulinum toxin for potential use as a weapon during World War II. Japan, Iraq, and the Soviet Union have also experimented with botulinum toxin as a biological weapon, and Iran, North Korea, and Syria may have done so as well. However, scientists have been perplexed and terrorists thwarted by how difficult it is to prepare botulism toxin for use as a weapon. Terrorists in Japan attempted to unleash it on at least three occasions between 1990 and 1995, but their plan for widespread destruction failed each time. In addition to tracking the efforts of bioterrorists, the U.S. government has developed elaborate methods to detect and respond to an attack with botulinum toxin. A national surveillance system involving doctors and hospitals has been designed to alert the U.S. Centers for Disease Control and Prevention to botulism outbreak and store of antitoxin are on hand to treat victims.

Types of botulism

There are three types of botulism:

- intestinal botulism
- food-borne botulism
- wound botulism.

Intestinal botulism

Intestinal botulism is the most common form of botulism. Children under the age of 12 months are most susceptible, but adults who have certain gastrointestinal problems may also be at risk. The incubation period for intestinal botulism is not known.

This form of botulism is caused by consuming food, particularly honey, or dust or soil that contains Clostridium botulinum spores, which then multiply inside the intestinal tract and produce toxins.

While honey has been implicated as a source of intestinal botulism in the United States, no Australian honey surveyed so far has contained the bacterium. Healthy adults aren't usually affected, because they have natural defences in their gastrointestinal tracts.

Food-borne botulism

While intestinal botulism involves consuming the botulism organisms themselves, food-borne botulism involves eating toxins formed by the bacteria in the food. These toxins are absorbed from the intestine, and act on the nervous system of the patient.

The symptoms of food-borne botulism are severe and without intensive care the poisoning is often fatal. Symptoms generally occur between 12 and 36 hours after eating the contaminated food.

Most human cases of botulism are caused by improperly cooked, preserved or canned foods where growth of the bacterium has produced the toxins.

Wound botulism

This rare form of botulism is caused by contaminated soil or gravel invading a wound. Intravenous drug users are also susceptible. Symptoms occur after four days to two weeks.

Occurrence of human pathogenic Clostridium botulinum among healthy dairy animals: an emerging public health hazard

The current study was conducted to investigate the occurrence of human pathogenic Clostridium botulinumin the feces of dairy animals. Fecal samples were collected from 203 apparently healthy dairy animals (50 cattle, 50 buffaloes, 52 sheep, 51 goats). Samples were cultured to recover C. botulinum while human pathogenic C. botulinum strains were identified after screening of all C. botulinum isolates for the presence of genes that encode toxins type A, B, E, F. The overall prevalence of C. botulinum was 18.7% whereas human pathogenic C. botulinum strains (only type A) were isolated from six animals at the rates of 2, 2, 5.8, and 2% for cattle, buffaloes, sheep, and goats, respectively. High fecal carriage rates of C. botulinumamong apparently healthy dairy animals especially type A alarm both veterinary and public health communities for a potential role which may be played by dairy animals in the epidemiology of such pathogen.

Botulism is a fatal neuroparalytic illness caused by botulinum neurotoxin (BoNT) produced by Clostridium botulinum. This neurotoxin is considered as the most lethal toxin known for humans at few nanogram quantities (25–50 ng). C. botulinum was firstly introduced to the scientific community in 1897 by van Ermengem who isolated such bacterium and identified it to be the cause of a severe paralytic illness outbreak that occurred in Belgium 1895; Bacillus botulinus. Afterward, the organism was assigned to the genus Clostridium and was named Clostridium botulinum. C. botulinum is a Gram positive spore-forming anaerobic bacterium which comprises four distinct phentotypic groups (I–IV), with groups I and II being pathogenic for humans. Furthermore, C. botulinum strains produce seven serologically different types of neurotoxins that were designated A–G, but types A, B, E, and F were the most implicated toxins in human outbreaks, while types C and D were associated with animal botulism. Botulism is usually associated with the ingestion of already preformed (BoNTs) in food and thereby the toxins enter the human body through gastrointestinal tract to produce the classical form of botulism which is called food-borne botulism. Nevertheless, other relatively uncommon forms of botulism were reported either when C. botulinumcolonized the intestinal tract of young infants and adults (infant botulism and adult intestinal colonization botulism) or when contaminated wounds become infected with such bacterium (wound botulism), germinating and producing neurotoxins

The clinical symptoms of botulism usually include double vision, dry mouth, dysphagia, dysphonia, paralysis of limb muscles which may rapidly extend to involve respiratory muscles, with subsequent respiratory failure and death. However, although food-borne botulism outbreaks seem to be relatively rare when compared with other causes of food poisoning, they show serious public health implications because of severe clinical outcomes and high treatment cost compared to other enteric infections. Food-borne botulism is usually associated with canned vegetables, meat, fish, and dairy products. Although, milk and milk products have not been frequently incriminated in food-borne botulism outbreaks, an increase in the number of outbreaks of botulism in dairy cattle was noted, alarming the public health community about the safety of milk and milk products. Notably, dairy products such as cheese and yogurt were previously traced as sources for several outbreaks of botulism. On the other side, much remains unknown about the occurrence of C. botulinum in the feces of apparently healthy dairy animals which may be considered as the main source of contamination of milk and milk products with C. botulinum spores. Therefore, the current study was carried out to investigate the occurrence of C. botulinum in the feces of apparently healthy dairy animals (cattle, buffalo, sheep, and goats) and clarify the actual role of these animals in the epidemiology of this infection, prevention, and control.

MATERIALS AND METHODS

Fecal samples were collected from the rectum of 203 dairy animals (50 cows, 50 buffaloes (Bubalus bubalis), 52 ewes, and 51 goats) using sterile gloves, inserted in sterile cups and transported directly to the laboratory in an ice box. All were lactating animals, apparently healthy and fed on pasture and concentrates (silage was not included in animal ration). Upon arrival, fecal samples were heated at 60 °C for 10 min then each was inoculated into two tubes of tryptose peptone glucose yeast extract (TPGY) broth. One tube was incubated at 37 °C, while the other was placed at 28 °C. All tubes were incubated for 5–7 days under anaerobic conditions using anaerobic jars and AnaeroGen™ gas-generating kits (Oxoid, UK). The enrichment cultures were further streaked on sheep blood agar plates and another incubation period was allowed for 3 days at duplicate temperatures, 28 and 37 °C as mentioned above under anaerobic conditions. Suspected colonies of C. botulinum showed narrow zones of beta hemolysis and were identified by Gram's stain, proteolytic activity in cooked meat media (Oxoid, UK), lipase test, catalase test, and other biochemical tests for confirmation.

The findings of the current study alarm both veterinary and public health communities about the occurrence of human pathogenic C. botulinum in the feces of apparently healthy dairy animals which is the initial source of such pathogen in dairy environment, milk, and milk products. So, it is to be noted that continuous surveillances are needed for dairy farms to determine the prevalence of C.

botulinum among the dairy herd while keeping strict hygienic measures during milking process. On the other hand, adequate quality control in dairy plants during heat treatment processes and also monitoring the cold chain in retails during storage of milk products should be applied in order to combat such deadly pathogen.

Causes

Botulism is rare, but there are still cases every year.

There are four ways to get botulism:

- Foodborne botulism occurs when food contaminated with the toxin is eaten. The bacteria are dead or gone by the time you eat the food, but the toxin remains. Most foodborne botulism is attributed to home-canned foods, but outbreaks occasionally occur in commercially-prepared foods.
- Infant botulism occurs when infants under 6 months of age eat bacterial spores rather than the toxins. Spores of C. botulinum then grow in the infant's intestines, where they produce toxins. Honey can contain spores of C. botulinum and is associated with infant botulism and should not be given to children under 6 months old. Infant botulism is rare.
- Adult intestinal colonization botulism occurswhen C. botulinumspores are ingested. The spores grow and eventually produce toxins in the digestive system. People with altered gastrointestinal systems and local microbes (due to surgery, presence of inflammatory bowel disease, or exposure to antibiotics) are most often affected.
- Wound botulism occurs when live bacteria infect an open cut and the toxin is carried through the body by the blood.

Rarely, toxins can be inhaled or absorbed through the eyes.

Ironically, some people intentionally get localized botulism from their doctor these days. The toxin is used to make a medication that is injected into twitching and spasmodic muscles to temporarily paralyze them. It's a treatment used for a variety of nerve disorders and for the cosmetic purpose of removing frown lines, forehead wrinkles, and "crow's feet" around the eyes.

Botulism can cause death due to paralysis of muscles used in breathing. However, advancements in supportive care have greatly reduced the death rate, and now less than 10% of people who get botulism die from the condition.

Symptoms and Complications

C. botulinum toxin attacks nerve endings, damaging them and then stopping muscle function. It can't cross the barrier that protects the brain. The heart has its own wiring system and also isn't affected. However, botulin can affect all the other nerves of the body, including the nerves of the muscles that operate the lungs. Botulin toxin kills by shutting down breathing.

Symptoms usually appear quickly in foodborne botulism, within 12 to 36 hours, but a few people feel no symptoms for as many as 8 days after eating the poison. Dry mouth, vomiting, nausea, diarrhea, and abdominal cramps are the first signals. Then the neurological symptoms begin to appear – the muscles slowly shut down, starting with the temple and forehead and proceeding slowly down both sides of the body. The face goes slack and expressionless, the eyelids droop, and the victim may drool. Vertigo and double vision are common. The arms slowly get weaker, then the legs. Diarrhea is replaced by constipation. By this time there's also difficulty in talking, swallowing, and breathing. Temperature and pulse remain normal.

The first sign of infant botulism is constipation. The neurological symptoms are the same as in foodborne botulism but develop less rapidly. This is because the spores produce toxin slowly and the child absorbs it bit by bit instead of all at once as in foodborne botulism. The baby may also suck milk weakly, have difficulty crying loudly, lose head control, and lack facial expression.

Making the Diagnosis

The Botulism Reference Service for Canada conducts the laboratory investigation for diagnosing botulism. The diagnosis of foodborne botulism is made by verifying whether the botulin toxin is in the food. Alternatively, C. botulinum can be isolated from the stool or stomach fluids. The diagnosis of intestinal botulism is made by identifying C.botulinum organisms and toxin in the feces. The diagnosis of wound botulism is made by detecting the toxin in the wound or by isolating C. botulinum from a positive wound culture.

Treatment and Prevention

Supportive care is provided to anyone with botulism. This may include hospitalization, use of a ventilator to help with breathing, and nursing care. Recovery can take many weeks as paralysis slowly improves.

While there is a botulism antitoxin and immune globin therapy, they are not approved for sale in Canada and are currently only available through Health Canada's Special Access Program.

The antitoxin works by neutralizing botulin. However, the antitoxin needs to be administered as soon as possible after diagnosis. The medication can't repair nerve endings already damaged. This means that while you may survive, you will be temporarily left in the state you were in at the time the medication was administered. That may mean paralysis, sometimes to the point of being unable to talk or even swallow.

The good news is that you'll recover, because new nerves can grow to replace those that were killed. However, this is a slow recovery process and it may take weeks or months, even up to a year, for this to occur.

The antitoxin is generally not given for infant botulism.

For wound botulism, antibiotics are given in addition to supportive care.

Preventing botulism is usually possible. However, if you prepare and store a lot of your own food, you may be creating unnecessary risk.The C. botulinum bacterium is anaerobic, meaning it likes airless environments. This is why you can get foodborne botulism out of a can. Very few cases of botulism arise from commercially canned food. Most occur when individuals can their own products.

Botulin toxin has been found in North America in these foods:

- asparagus
- baked potatoes
- beets
- canned corn
- chicken, chicken livers, and liver paté
- garlic-in-oil dressing
- green beans
- ham
- lobster
- luncheon meats
- mushrooms
- peppers
- ripe olives
- sausage
- smoked and salted fish
- soups
- spinach

- stuffed eggplant
- tuna fish

Only highly acidic foods are safe from C. botulinum. Freezing will shut down toxin production, but a fridge isn't cold enough. The following food handling procedures can help you to prevent foodborne botulism:

- Preserved food should be heated to a temperature above 100°C (212°F) for at least 10 minutes during the canning process.
- Cook food at 79.9°C (176°F) for at least 30 minutes. This usually destroys toxins.
- Do not eat or store cooked foods that have been at room temperature for 4 hours or more.
- Do not eat foil-wrapped baked potatoes that have been left at room temperature, and do not store chopped garlic or onions in oil at room temperature.
- If eating home-canned food, boil it first with frequent stirring for 10 minutes.

Other susceptible species

Botulism can occur in many vertebrates and invertebrates. Botulism has been reported in such species as rats, mice, chicken, frogs, toads, goldfish, aplysia, squid, crayfish, drosophila and leeches.

Death from botulism is common in waterfowl; an estimated 10,000 to 100,000 birds die of botulism annually. The disease is commonly called "limberneck". In some large outbreaks, a million or more birds may die. Ducks appear to be affected most often. An enzootic form of duck botulism in Western USA and Canada is known as "western duck sickness". Botulism also affects commercially raised poultry. In chickens, the mortality rate varies from a few birds to 40% of the flock.

Botulism seems to be relatively uncommon in domestic mammals; however, in some parts of the world, epidemics with up to 65% mortality are seen in cattle. The prognosis is poor in large animals that are recumbent.

In cattle, the symptoms may include drooling, restlessness, uncoordination, urine retention, dysphagia, and sternal recumbency. Laterally recumbent animals are usually very close to death. In sheep, the symptoms may include drooling, a serous nasal discharge, stiffness, and incoordination. Abdominal respiration may be observed and the tail may switch on the side. As the disease progresses, the limbs may become paralyzed and death may occur. Phosphorus-deficient cattle, especially in southern Africa, are inclined to ingest bones and carrion containing clostridial toxins and consequently develop lame sickness or lamsiekte.

The clinical signs in horses are similar to cattle. The muscle paralysis is progressive; it usually begins at the hindquarters and gradually moves to the front limbs, neck, and head. Death generally occurs 24 to 72 hours after initial symptoms and results from respiratory paralysis. Some foals are found dead without other clinical signs.

Clostridium botulinum type C toxin has been incriminated as the cause of grass sickness, a condition in horses which occurs in rainy and hot summers in Northern Europe. The main symptom is pharynx paralysis.

Domestic dogs may develop systemic toxemia after consuming C. botulinum type C exotoxin or spores within bird carcasses or other infected meat but are generally resistant to the more severe effects of Clostridium botulinum type C. Symptoms include flaccid muscle paralysis. Muscle paralysis can lead to death due to cardiac and respiratory arrest.

Pigs are relatively resistant to botulism. Reported symptoms include anorexia, refusal to drink, vomiting, pupillary dilation, and muscle paralysis.

In poultry and wild birds, flaccid paralysis is usually seen in the legs, wings, neck and eyelids. Broiler chickens with the toxicoinfectious form may also have diarrhea with excess urates.

7

CAPRINE ARTHRITIS ENCEPHALITIS: A VIRAL DISEASE

Caprine arthritis and encephalitis (CAE) is a viral disease of goats characterised by arthritis mainly in the carpel joints in mature goats (big knee), and by encephalitis in goat kids. Infected goats can also suffer from pneumonia, mastitis and chronic wasting. The infection is lifelong, untreatable and significantly affects productivity.

Caprine arthritis encephalitis (CAE) is a viral disease of goats caused by a lentivirus called caprine arthritis encephalitis virus. The disease is found worldwide.

Two syndromes of CAE occur. Adult goats develop a chronic progressive arthritis, whereas young goats develop a neurological syndrome, with signs of paresis or paralysis. Less commonly, mastitis or pneumonia may occur.

Infection is life-long, and it may be years before signs of the disease occur. The reason for the long (and variable) period of dormancy of the virus is not known.

In goats which develop arthritis, the joints become inflamed and swollen, and the goats will slowly lose condition. In some cases the goat will not be able to stand.

In goats which develop the neurological form of the disease, the onset of signs is gradual over several weeks. The hind legs are most often affected. The goat will be uncoordinated, and unable to place its feet properly, so that it "knuckles", that is, it stands with the front of its fetlock on the ground, rather than its hoof. The goat has increased difficulty standing and eventually is unable to stand.

The disease is spread to goat kids when they drink colostrum or milk from infected goats. Separating goat kids from infected goats, and feeding the kids with cow's milk, or pasteurized goat milk, will prevent infection. The disease can be spread from goat to goat via direct contact and body fluids, such as saliva. Blood testing goats for CAE virus before moving them into a new herd will prevent the spread of the disease.

There is no known cure. To prevent spread of the disease, infected animals are separated from non-infected goats, or culled.

CAUSE

A retrovirus belonging to the subfamily Lentvirus.

Other names

Caprine retrovirus

CRV

Big knee in goats

Distribution

Present in Queensland, mainly in dairy goats in eastern parts of the state. It has also been reported in the United States, Canada, Europe and New Zealand.

Life cycle

Most goats are infected when young through colostrum or milk from infected does. They generally develop the disease months or years after being infected. They remain infected for life.

Affected animals

goats

Clinical signs

The signs of disease vary according to the age of the animal. Effects can last several months.

Young goats

Symptoms in young kids may include:

Encephalitis in young kids 2-4 months old a short, stilted gait occurring suddenly, and progressing to arthritis and paralysis pneumonia hyperaesthesia developing, with convulsions and death.

Affected kids often retain their alertness and appetite during the early stages of the disease.

Mature goats

Symptoms in mature goats include:

depression

weight loss

changes to coat (becoming more rough)

mastitis

enlarged and painful joints, leading to lameness, incoordination and progressive paresis and paralysis

arthritis, generally in mature goats around 1-2 years old (but can occur as early as 4 months).

Some mature goats may also suffer from encephalitis.

Impacts

Economic

Economic impacts may include:

high cull rates

reproductive losses

milk production losses

animal deaths

loss of trade.

HOW IT IS SPREAD

Most goats are infected when young through infected colostrum or milk from infected does. Once infected, the animal remains infected for life. Horizontal infection in a herd is possible through infected items at feeding and watering points etc.

Breeding and in utero transmission to fetuses are not considered risks because the virus does not cross the placenta and infected does give birth to normal kids. The birthing process can be a risk factor for infection if vaginal tears expose kids to blood.

People with contaminated hands, clothing and footwear can spread the infection into a herd.

Infection can happen at all stages of life, but kids are most at risk. The disease is generally slow to develop and infected animals could take months or years to show clinical signs.

RISK PERIOD

Infection can happen at all stages of life, but kids are most at risk.

Monitoring and action

Blood (serum) samples are required to diagnose CRV by an ELISA test or AGID test.

Only introduce goats from CAE-free herds, or have their blood tested before joining them to herd.

Remove all infected goats from the herd.

Remove kids from infected does at birth and keep them in isolation. Only feed them bovine colostrum and milk, milk substitutes, or milk from known CAE-free does. Kids affected by the disease should be humanely destroyed.

There is no successful treatment for infected animals.

CAE scheme requirements include:

flock management and biosecurity

2 negative blood tests of all goats over 6 months of age for initial accreditation (1 negative test if goats are purchased from an accredited herd) annual or biennial re-accreditation (depending on the time of initial accreditation).

If the flock becomes infected, it will no longer be accredited until all reactor animals are removed from the herd and the flock is tested back to accreditation standards.

Diagnosis of Caprine Arthritis and Encephalitis

Based on clinical signs and history

Serologic testing is useful to determine herd CAEV status

A presumptive diagnosis of caprine arthritis and encephalitis can be based on clinical signs and patient history. Infectious arthritis caused by Mycoplasma spp and traumatic arthritis are differential diagnoses for arthritis due to CAEV infection. Differential diagnoses for the progressive paresis and paralysis exhibited by young kids should include enzootic ataxia, spinal cord abscess, cerebrospinal nematodiasis, spinal cord trauma, and congenital anomalies of the spinal cord and vertebral column. If a neurologic examination indicates brain involvement, polioencephalomalacia, listeriosis, and rabies should be considered. The pulmonary form of caseous lymphadenitis may have a similar clinical presentation to the pulmonary form of CAEV in adult goats.

Both an agar gel immunodiffusion test and ELISA for CAEV are considered sufficiently reliable for use in herd control programs. The agar gel immunodiffusion test is reported to be more specific but less sensitive than the ELISA test. A positive test result in an adult goat implies infection but does not confirm that the clinical signs are caused by CAEV. Kids infected at birth develop a measurable antibody response 4–10 weeks after infection. However, positive test results in kids <90 days old usually reflect colostral antibody transfer. Negative test results do not reliably exclude CAEV infection, because the time for postinfection seroconversion is variable, and some goats have a very low antibody titer that may not be detectable. Low antibody titers are common in late pregnancy. Because of the limitations of serologic testing, definitive diagnosis of clinical CAEV requires demonstration of characteristic lesions in biopsy specimens or post-mortem. Virus isolation or PCR assay to demonstrate presence of viral antigen in tissues may be used to further confirm the diagnosis.

SUPPORTIVE CARE IS INDICATED

Appropriate management practices, particularly of kids (eg, avoiding the use of pooled colostrum), may help control disease transmission There are no specific treatments for any of the clinical manifestations of caprine arthritis and encephalitis virus infection. However, supportive treatments may benefit some patients. The condition of goats with polysynovitis-arthritis may be improved with regular foot trimming, use of additional bedding, and administration of NSAIDs such as meloxicam, flunixin meglumine, phenylbutazone, or aspirin. Goats with encephalomyelitis may be managed for weeks with good nursing care. Antimicrobial treatment is indicated to treat secondary bacterial infections that may complicate interstitial pneumonia or indurative mastitis components in cases of CAEV infection. Providing high-quality, readily digestible feed to goats positive for CAEV may delay the onset of wasting. Eventually, euthanasia may be necessary for animals with advanced disease.

In commercial herds, one or more of the following have been recommended for control of CAE: 1) permanent isolation of kids beginning at birth; 2) feeding of heat-treated colostrum (45°C [113°F] for 60 minutes) and pasteurized milk; 3) semi-annual serologic testing of the herd, with identification and segregation of seronegative and seropositive goats; and 4) eventual culling of seropositive goats. If the control program includes segregation of herds into seropositive and seronegative groups, groups should be separated by a minimum of 6 ft (1.8 m), and shared equipment should be disinfected using phenolic or quaternary ammonium compounds.

Clinical Findings for Caprine Arthritis and Encephalitis

Clinical signs of caprine arthritis and encephalitis may be observed in ~20% of CAEV–infected goats during their lifetime. The most common manifestation of infection is polysynovitis-arthritis, which is seen primarily in adult goats but can occur in kids as young as 6 months old. Signs of polysynovitis-arthritis include joint swelling and lameness of varying severity. The carpal joints are most frequently involved. The onset of arthritis may be acute or chronic, but the clinical course is always progressive. Affected goats also lose weight and usually have poor hair coats.

Encephalomyelitis is generally seen in kids 2–6 months old but has been reported in older kids and adult goats. Affected kids initially exhibit weakness, ataxia, and hind limb proprioceptive deficits. Hypertonia and hyperreflexia are also common. Over time, signs progress to paraparesis or tetraparesis and paralysis. Depression, head tilt, circling, opisthotonos, torticollis, and paddling of the limbs have also been reported.

Interstitial pneumonia due to CAEV infection rarely produces clinical signs in kids. However, in adult goats with serologic evidence of CAEV infection, chronic interstitial pneumonia may lead to progressive dyspnea. The "hard udder" syndrome attributed to CAEV infection is characterized by a firm, swollen mammary gland and agalactia at the time of parturition. Milk quality is usually unaffected. Although the mammary gland may soften and produce close to a normal volume of milk, production remains low in many goats with indurative mastitis.

Lesions

Histopathologic lesions due to CAEV infection are generally described as lymphoproliferative, with degenerative mononuclear cell infiltration. Lesions in joints are characterized by thickening of the joint capsule and marked proliferation of synovial villi. In chronic cases, soft-tissue calcification involving joint capsules, tendon sheaths, and bursae is not uncommon. Severe cartilage destruction, rupture of ligaments and tendons, and periarticular osteophyte formation have also been described in advanced cases.

Gross lesions associated with the neurologic form of CAEV include asymmetric, brownish pink, swollen areas, most commonly located in the cervical and lumbosacral spinal cord segments. Histopathologically, these lesions are characterized by multifocal, mononuclear cell inflammatory infiltrates and a varying extent of demyelination. During gross examination, lungs of affected goats may be seen as firm and gray-pink, with multiple, small, white foci, and do not collapse.

The bronchial lymph nodes are invariably enlarged. Microscopic features of articular lesions include synovial cell hyperplasia, subsynovial mononuclear cell infiltration, villous hypertrophy, synovial edema, and synovial necrosis. Histopathologic findings include chronic interstitial pneumonia with mononuclear cell infiltration in alveolar septae and in perivascular and peribronchial regions. In does with udder induration, mononuclear infiltration of periductular stroma obliterates healthy mammary tissue.

8

AKABANE DISEASE

Akabane disease, an infectious disorder causing congenital abnormalities in calves, was studied in naturally affected calves between 1972 and 1974 in Japan. The animals included one three-month fetus from which a strain of Akabane virus (OBE-1) was isolated, and a total of 177 stillborn or premature fetuses and deformed or infirm newborn calves that died within a few days of birth. The three-month fetus had nonpurulent encephalomyelitis in the undifferentiated central nervous system, characterized by necrosis of the nerve tissue and endothelial proliferation in the encephalitic process; and polymyositis in the myotubule phase with parenchymal degeneration and cellular infiltrates in the skeletal muscle.

The full-term fetuses and newborn calves had nonpurulent encephalomyelitis in the early stage of the epizootic. In the middle to late stages, a dysplastic muscular change which we call runt-muscle disease was seen simultaneously with a decrease in number of ventral horn neurons in the spinal cord and arthrogryposis in the legs. Cystic cavities and thick vascular walls sometimes were seen in the central nervous system in these stages. Hydranencephaly was prominent in the late stage.

Akabane virus (AKA virus) is a Culicoides transmitted teratogenic virus, which infects bovine, ovine and caprine fetuses in utero causing congenital arthrogryposis and hydranencephaly (A-H) and, less frequently, congenital polioencephalomyelitis and rarely, acute encephalitis in both young calves and adult cattle (Walton, 1992). It is one of about 25, related Simbu serogroup viruses (Family: Bunyaviridae, genus orthobunyavirus) including previously unrecognized viruses such as Schmallenberg virus in Europe (Hoffmann et al.,

2012). Only a small number of the Simbu viruses have been associated with disease in animals.

Akabane virus has a relatively narrow host distribution. Antibody to AKA virus is extremely common in cattle, buffalo (Syncerus caffer and Bubalus bubalis), camels and small domestic ruminants in many tropical and temperate regions throughout the world. Horses, zebra and many of the wild ruminant species in Africa (wildebeest, kongoni, waterbuck, topi, eland, bushbuck, Impala, Grants gazelle and Thomson's gazelle), show evidence of infection with AKA virus. Throughout much of its range however, where most cattle become infected with AKA before they reach breeding age, there is no evidence of any pathology in the known hosts.

There is evidence to show that AKA virus activity occurs in a wide range of tropical, subtropical and temperate ecological zones in Africa and Asia as determined by the availability of suitable habitat for its insect vectors. Thus, although it may have only been studied in a small number of countries, the ecology of the virus would suggest that it is probably more widespread than recognised. As AKA virus is predominantly a fetal pathogen, disease is not an indicator for its distribution due to high levels of immunity in animals prior to reaching reproductive maturity. Serological surveys are required to identify its distribution.

The clinical congenital A-H syndrome due to AKA virus has a limited geographical distribution. Outbreaks are most commonly observed at the limits of the range of the vector, where periodic extensions occur, beyond where it is known to be endemic and where the ruminant hosts are susceptible or where the vectors are usually scarce or have only a brief and irregular period of seasonal activity. Reduction in the vector distribution in endemic areas under the influence of adverse climatic conditions such as drought can create susceptible generations of cattle that experience disease when the usual level of vector activity resumes. Disease is also observed when susceptible pregnant animals are introduced into endemic areas during the vector season.

Akabane disease causes reproductive failure and foetal deformities in cattle, sheep and goats.

Scientific name

Akabane disease

CAUSE

Akabane virus is an arbovirus and a member of the Simbu serogroup in the family Peribunyaviridae, genus Orthobunyavirus, family Bunyavirus.

DISTRIBUTION

Akabane virus is common in many tropical and subtropical areas in the Eastern Hemisphere, including parts of Asia, Africa, the Middle East and Australia. Disease has been recorded in Japan, Korea, Taiwan, Australia, Israel and Turkey.

The distribution of Akabane virus in Australia depends on the seasonal distribution of Culicoides brevitarsis (a biting midge) that transmits the virus.

Within Australia, Akabane virus occurs in the Kimberley region of Western Australia, the top half of the Northern Territory and throughout most of Queensland. With seasonal variation, Aakabane virus also extends to the far southern eastern seaboard of New South Wales and adjacent hinterlands.

CONTROL

The main form of prophylaxis is vaccination. Inactivated vaccines have been developed for Akabane virus cultured on hamster lung cell cultures and these are used in Japan and Australia (Anon., 1999a; Inaba and Matumoto, 1990). The time of vaccination should be geared to ensure maximum protection when the peak incidence of the arthropod vectors is expected. In the tropical sub-tropical global belt, in late spring or early summer is probably optimal. A live vaccine has also been developed (Inaba and Matumoto, 1990). A novel trivalent vaccine for the teratogenic Aino, Akabane and Chuzan viruses has been developed in Korea (Kim et al., 2011). Due to the very high attack rates of Culicoides, while vector control by the use of insecticides and habitat modification may reduce insect numbers, it is not considered that these are sufficiently effective to prevent the occurrence of disease.

Movement of female breed stock into an endemic area at least 2 months prior to mating is likely to lead to protective immunity being established prior to pregnancy. The time required for most animals to develop immunity will vary depending on the number of infector vectors present.

Once disease is observed, remedial action is ineffective as all remaining foetuses are likely to be infected.

There is no treatment for Akabane affected offspring.

CLINICAL DIAGNOSIS

Akabane infection may be suspected when there is an increase in cases of neurological disease in newborn calves in late summer-autumn. Although adult animals do not normally exhibit clinical signs, the Japanese Iriki strain and some other virulent AKAV strains may cause acute encephalitis following postnatal infection, even at times in adult animals (Lee et al., 2002; Kamata et al., 2009; Oem at al., 2012; Zentis et al., 2012). Cases of acute encephalitis may soon be followed by a sudden increase in the rate of abortions and large numbers of aborted, mummified, premature or stillborn fetuses with AG and HE or other malformations. Fetuses are badly deformed, many dead at birth and in AG cases, the limbs are locked in position. Infected live-born young may exhibit CNS and muscular degeneration, which prevents them from standing and suckling.

Calves with HE may survive for some months if hand-reared. AG may be accompanied by torticollis, cervical scoliosis, brachygnathism and kyphosis. Lesions of HE are expressed clinically as blindness, nystagmus, microphthalmia, deafness, dullness, slow suckling or dysphagia, paralysis and ataxia. The presence of gross lesions in the CNS can aid in diagnosis while histological changes in the motor neurones of the spinal cord are suggestive. There may also be muscular lesions.

The seasonal clustering of the concurrent birth of large numbers of sheep or goats with congenital defects and with highly suggestive gross pathology also increases suspicion. These cases would be expected to occur during the late winter and spring months.

DIFFERENTIAL DIAGNOSIS

When there is an outbreak of abortions and still births involving congenitally deformed ruminants, apart from AKAV, other vector borne viruses that must be excluded are other orthobunyaviruses such as Aino and Schmallenberg. When there is a small number of cases, other viral infections such as bovine viral diarrhoea virus (BVDV) and genetic diseases should also be excluded. In cattle, the cerebellum is rarely, if ever, affected following AKAV infection, a useful differential feature to distinguish from other congenital infections such as BVDV.

Similar fetal abnormalities in sheep or goats may be caused by vaccination with a modified live vaccine strain (the Smithburn neurotropic strain) of Rift Valley fever virus, which is widely used as a vaccine in Africa. However, there are not usually as many cases as seen in epidemics of AKA disease. A history of vaccination during the first trimester of pregnancy should prompt investigations as to whether a vaccine may have been the cause. The differential diagnosis for neurological signs or behavioural changes can also include sheep associated malignant catarrhal fever (SA-MCF), transmissible spongiform encephalitis (TSE) and listeriosis.

LABORATORY DIAGNOSIS

Confirmation of AKA virus infection must be guided by an understanding of the pathogenesis of this disease and the biology of AKAV infection of the dam. As a disease predominantly of the fetus, the emphasis is on confirmation of infection in unsuckled ruminant neonates, stillborn or aborted fetuses. The capacity of the ruminant neonate to develop a strong immune response by full term provides an important approach to diagnosis. Infected bovine fetuses develop antibodies at from about 70 days of gestation (Inaba and Matumoto, 1990).

The detection of AKAV specific antibody in either serum or body fluids (e.g. pericardial or pleural) is considered to be diagnostic. To guide such an approach, prior to testing for virus-specific antibodies, there is considerable value in testing for total IgG levels in these fluids. An elevated level of IgG can then be followed by specific antibody tests for AKAV or for other viruses (e.g. for other orthobunya viruses or BVDV). There are several different serological tests available to detect AKAV specific antibodies but those most frequently used are the VNT and ELISA. Tests such as the AGID and IFAT are more broadly reactive and have lower sensitivity but can still detect fetal antibodies to AKAV and closely related viruses.

In the absence of an elevated IgG level, efforts can be directed at virus detection by either virus isolation or PCR. However, the possibility that the case is not due to an infectious aetiology should be first considered and excluded. Virus detection is also much more likely to be successful in fetuses that are aborted in mid-gestation and soon after infection. Although virus isolation can be successful, there are considerable advantages in using PCR due to the higher sensitivity and the capacity to detect residual RNA that may be present in fetal tissues long after the virus is no longer infectious. Both conventional and real time PCR assays are available (Akashi et al., 1999; Stram et al., 2004b). Although viral RNA can be detected in a range of internal organs (especially spleen, lymph nodes and thymus), the brain and spinal cord remain tissues of choice. Residual RNA can also be detected from fetal fluids and the placentome.

Virus isolation can be attempted from the placenta (caruncle) of the dam and fetal membranes (placentome) or chorio allantoic fluids, fetal blood, muscle or nervous tissue of affected fetuses, especially when aborted. Virus isolation may be attempted using a range of cell cultures but those derived from hamsters (eg BHK21, Hm Lu-1) and monkeys (Vero, CV-1) are preferred. There can also be an increased isolation rate by first inoculating insect cell cultures (eg C6/36 cells) with tissue homogenates and fluids.

Due to the extremely short duration of viraemia in the dam (lasting about 3-5 days) there is little point in attempting virus detection on maternal blood or on animals thought to have been infected postnatally and are showing signs of an acute encephalitis. Adult animals rapidly produce antibodies to AKAV and are invariably seropositive by about 14 days after infection and will persist for several years in the absence of further exposure. Therefore maternal serology is of little use to confirm fetal infections due to AKAV but a negative result is convincing for exclusion purposes.

PATHOLOGY

The range of gross and histopathological changes that are observed in ruminants vary with the species (related to the length of gestation) and the time of gestation at which infection occurs. In cattle, there is a distinct suggestion of individual changes while in sheep and goats, multiple abnormalities can be observed in the same animal.

Calves that are infected late in gestation or sometimes soon after birth show acute neurological signs including a flaccid paralysis and have histological lesions of a non-suppurative polio-encephalomyelitis (Hartley et al., 1977; Kirkland et al., 1988).

In calves with arthrogryposis (AG), following infection between 5-6 months of gestation, apart from the fixation or severely restricted range of movement of joints, there are few other grossly detectable changes. These changes are largely due to degenerative changes in muscles and tendons arising from a neurogenic dystrophy. Microscopically there are severe degenerative changes in the motor horns of the spinal cord (Hartley et al., 1977; Kirkland et al., 1988). In some cases, degenerative changes are also apparent in the skeletal muscle.

When calves are stillborn or show behavioural changes, grossly apparent defects are likely to be apparent in the brain. These changes are the outcome of infection in the 3rd and 4th months of pregnancy and can vary from small cystic lesions to the virtual absence of the cerebral hemispheres and replacement with fluid filled meningeal sacs (hydranencephaly; HE) (Whittem, 1957; Shepherd et al., 1978; Hartley et al., 1997; Kirkland et al., 1988). There are few microscopic changes and histopathology is of minimal diagnostic value. There will be an absence of large areas of brain surrounded by tissues with relatively normal architecture. In cattle, the cerebellum is rarely, if ever, affected.

If a fresh aborted fetus is found, depending on the age of the fetus and time since it was infected, gross lesions may not be apparent. A range of acute necrotic, degenerative changes and perhaps a mild to moderate non-suppurative encephalomyelitis suggestive of a viral infection may be detected. The lesions can be detected in all parts of the CNS, with perivascular cuffing, neuronal degeneration and cavitation of the brain and neuronal degeneration in the motor neurones of the spinal cord. Muscular dystrophy may also be observed.

In sheep and goats, there can be varying combinations of gross and histological lesions observed in the same animal. As well as the abnormalities of the central nervous and musculoskeletal systems, developmental changes may sometimes be observed and include hypoplasia of the lungs and thymus.

9

EDWARDSIELLA TARDA

Edwardsiellosis, caused by Edwardsiella tarda, has been reported worldwide in economically important fish species, including Japanese eel (Anguilla japonica), red sea bream (Pagrus major), yellowtail (Seriola quinqueradiata), channel catfish (Ictalurus punctatus), and turbot (Scophthalmus maximus). This infection also leads to serious economic losses in the aquaculture of olive flounder (Japanese flounder; Paralichthys olivaceus), the most important fish species in South Korean aquaculture, with production valued at 489.7 billion Korean Won (40 922 MT), which corresponds to 56.5% of total fisheries production in 2010.

Recent studies on vaccine development have applied a variety of antigen-preparation methods; however, commercial vaccines are not yet available. In addition, numerous studies have reported on virulence factors of E. tarda and immune responses of hosts. In the present study, the pathogenicities of E. tarda in fish that can be exploited to elicit effective protection strategies against edwardsiellosis will be discussed.

Edwardsiella tarda, while a common pathogen in some cultured fish species, only occasionally infects cultured striped bass. Herman and Bullock (1986) reported that E. tarda infected 4 to 5 cm fingerling striped bass cultured in freshwater. The fish became moribund, swam lethargically at the surface, displayed pale gills and had a slightly discolored area in the cranium. A pure culture of a gram-negative, cytochrome oxidase negative, motile rod, presumptively identified as Edwardsiella tarda was isolated from lateral line and kidney tissues.

Additional biochemical characteristics include production of indole from tryptone and an alkaline slant over an acid butt in triple sugar iron agar with production of H2S and gas. Naive striped bass were infected with this organism by water borne exposure. Histologically, the epithelium of experimentally infected fish was necrotic and the fins were frayed. Numerous abscesses were present in the kidney along with necrosis of the hematopoietic tissue of the trunk kidney. As the disease progressed, the kidney became grossly enlarged with massive inflammation, necrosis and a large number of bacteria being present. The lesion expanded to the adjacent trunk musculature.

EDWARDSIELLOSIS

Definition

Edwarsiellosis is acute or subacute or chronic bacterial disease of a variety of fish species caused by Edwardsiella spp. Characterized by septicaemic picture and abscess formation.

Causative agent

The genus Edwardsiella includes two species of bacteria that cause major disease in fish. Edwardsiella tarda infects fish and other animals while E. ictaluri infects fish only. Species of Edwardsiella, are a member of the Enterobacteriaceae, are small & straight rods. They are gram negative, usually motile and facultatively anaerobic.

They are non-spore-forming, non-capsulated & cytochrome oxidase negative. Similarities between E. tarda and E. ictaluri have been observed except that isolates of E. ictaluri did not produce H2S or Indole or ferment glucose with the production of gas at 37úC. Moreover the isolates did not agglutinate with antiserum of E. tarda. Both bacteria isolated form internal organs, intestine and muscular lesions by inoculation of material on brain heart infusion agar or blood agar. The optimum growth temperature range form 26 to 30úc. The colonies are small, round, convex and transparent within 24-48h.

Susceptible species

E. tarda infects a wide variety of fish. The most predominant hosts are eel and catfish, but many other fish groups are also susceptible. At least 21 species of fish are known to have been infected and probably all species of fish are susceptible under certain conditions. E. tarda infections are not limited to fish, but often exist as part of the normal intestinal microflora especially in fish eating birds, reptiles, cattle and swine. Warm-blooded animals including humans have been known to be susceptible to E. tarda.

Stress factors:

1. Fluctuating and high temperature of the water.
2. Low dissolved oxygen.
3. High organic content.
4. Overstocking.
5. High ammonia in water.
6. Rough handling of fish.
7. Improper diet.
8. Low water chlorides.

Mode of the transmission:

1. The mode of transmission may occur from fish to other fish via the water by organisms shed with the feces.
2. Cannibalism of infected fish.
3. Feeding of dead or infected carcasses.
4. Birds help in spreading of infection by picking up dead fish from one pond and flying to another pond and dropping the infected carcasses to healthy one.
5. Contaminated nets and equipment can transfer the disease from pond to another.
6. Carriers help in spreading of the disease.
7. Edwardsiella species inhabit in the aquatic invertebrates and human beings. Any one of those hosts may serve as reservoir.
8. The micro-organism found normal in the intestinal tract and when fishes put under any stress condition the bacterium have ability to invade the fish and causes the infection.

Clinical signs:

1. Clinical signs of E. ictaluri infection includes:

a. Sudden decrease in feeding (change in feeding activity).

b. There are two behaviors that are very characteristic of fish infected with E. ictaluri. These are spiraling and surfing (unusual swimming patterns of infected fish may be caused by bacterial infection in the brain).

c. Affected fish also sometimes hang in the water column with head up and tail down.

d. Open lesions found on the head of diseased fish particularly on the frontal bone of the skull between the eyes (hole in head) usually found in fingerling catfish and not seen in older one.

e. White or hemorrhagic spots found in the skin.

f. Ascitis.

g. Exophthalalmia.

h. Pale colouration of skin and gills.

i. Areas of hemorrhage may be observed at base of fins, around the head, operculum and abdomen.

j. Clear or bloody fluid is often present in body cavity of fish.

k. Spleen, kidney may be soft, pale in color and have bloody spots.

l. Liver has pale areas of tissue destruction (necrosis).

m. Petechial hemorrhages can be found in the muscles, intestine and fat of the fish.

n. The intestine is also often filled with a bloody fluid.

CLINICAL SIGNS OF E. TARDA INFECTION

Clinical manifestation of E. tarda in mild infections is presence of small cutaneous lesions of approximately 3-5 cm in diameter which are located in the posterior lateral region of the body with progression of the disease, abscesses develop in the muscle of the body and tail. These abscesses may become large and develop into gas filled hollow areas. From the surface, these are seen poorly pigmented, convex and swollen areas which if punctured emit a foul odour. This condition has given rise to the name emphysematous putrefactive disease of catfish. Diseased in tilapia characterized by loss of pigmentation. Swollen of the abdomen due to accumulation of ascitic fluid. Protruded, haemorrhagic anus. Small white nodules may be observed in the gills, kidney, liver and spleen. These nodules are packed with bacteria.

Diagnosis

1. Case history.
2. Clinical signs and P.M.
3. Isolation and identification of the bacterium.

4. Fluorescent antibody or ELISA-test can be used for rapid diagnosis of the disease.

Treatment and control:

The disease can be treated with oxytetracycline 50 mg/kg of body weigh/day for ten days. To control the infection, following steps must be taken:

1. Stress factors must be removed.
2. Removal of sick and dead fish as soon as possible.
3. Preventing infected animals from coming into contact with the fish farm.
4. Good management of fish farm.

Zoonotic importance

E. tarda is responsible for several clinical conditions in humans such as diarrhoea gastroenteritis, typhoid-like illness, peritonitis with sepsis and induced abscesses in liver.

Edwardsiella tarda infects a wide variety of fish. The most prominent hosts are Japanese eel (Anguilla japonica) in Japan (Egusa, 1976) and channel catfish (Ictalurus punctatus) in the USA (Meyer and Bullock, 1973). However, many other fish groups are also susceptible; for example Japanese flounder (Paralichthys olivaceus) in Japan (Nakatsugawa, 1983), and tilapia (Oreochromis spp.) are significantly affected and isolations from tropical ornamentals have been reported (Vandepitte et al., 1983; Bullock and McCraran, 1989; Dixon and Contreras, 1992). Most diseases associated with E. tarda occur in cultured fish in either fresh or marine waters; at least 30 species of fish are known to have been infected and probably all species are susceptible under certain conditions. Edwardsiella tarda generally causes disease only in warm-water fish, but infections have also occurred in migrating chinook salmon (Oncorhynchus tshawytscha) in the Rogue River, Oregon, USA (Amandi et al., 1982), spawning Atlantic salmon (Salmo salar) in Nova Scotia, Canada (Martin, 1984), farmed rainbow trout (Onchorhynchus gairdneri) in Australia (Reddacliff et al., 1996) and brook trout (Salvelinus fontinalis) broodstock in Quebec, Canada (Uhland et al., 2000). Other marine species such as turbot (Scophthalmus masimus) (Nougayrede et al., 1994) and sea bass (Dicentrarchus labrax) (Blanch et al., 1990) have also been reported to be infected with E. tarda.

Edwardsiella tarda infections are not limited to fish and in some instances other animals may simply be carriers and/or reservoirs of the pathogen. E. tarda can be found in reptiles (snakes, turtles and alligators) (Wallace et al., 1966; Nagel et al., 1982; Sugita and Deguchi, 1983), birds (White et al., 1969, 1973; Berg and Anderson,

1972), cattle (Ewing et al., 1965), swine (Arambulo et al., 1967), marine mammals (Coles et al., 1978) and other warm-blooded animals (Van Damme and Vandepitte, 1984). In many instances, E. tarda is part of the normal intestinal microflora of aquatic animals (Wyatt et al., 1979; Van Damme and Vandepitte, 1980; Kanai et al., 1988). Bauwens et al. (1983) found E. tarda in the intestine of a variety of zoo animals, especially fish-eaters and water-loving species. Edwardsiella tarda has also been associated with several different manifestations in humans (King and Adler, 1964; Jordan and Hadley, 1969; Bockemuhl et al., 1971; Van Damme and Vandepitte, 1980,1984; Vandepitte et al.,1983; Janda and Abott, 1993b).

Although environmental stressors are not essential precursors to E. tarda infections in fish, high temperature, poor water quality and high organic fertility probably contribute to the onset and severity of the disease. Juvenile channel catfish that were experimentally infected with Aeromonas hydrophila and then exposed to environmental stressors (low dissolved oxygen, high ammonia and high carbon dioxide) developed E. tarda infections in 25-50% of the fish (Walters and Plumb, 1980). This compared to 4.5-12.5% E. tarda infections in the non-A. hydrophila injected or otherwise non-stressed fish. Indications were that environmentally induced stress and other bacterial infections could predispose channel catfish to endemic E. tarda. Cultured tilapia under stress are also susceptible to E. tarda (Benli and Yildiz, 2004). On several occasions, intensively cultured and environmentally stressed Nile tilapia (Oreochromis niloticus) in recirculating systems that had a moderate to heavy Trichodina (protozoan parasite) infestation, were also infected with E. tarda and Streptococcus spp. While stressed, neither bacterial infection responded to chemotherapy, but, as soon as the stressor was relieved and parasites were eliminated, the E. tarda infection disappeared. Many epizootics of E. tarda occur in fish predisposed to fluctuating water temperatures (Liu and Tsai, 1980; Amandi et al., 1982) or fish in highly enriched waters (Meyer and Bullock, 1973; Baya et al., 1997; Uhland et al., 2000). Other fish species susceptible by stress are striped bass (Morone saxatilis) (Herman and Bullock, 1986; Baya et al., 1997) and rainbow trout (Redacliff et al., 1996).

Pathogenesis of and strategies for preventing Edwardsiella tarda infection in fish

Edwardsiella tarda is one of the serious fish pathogens, infecting both cultured and wild fish species. Research on edwardsiellosis has revealed that E. tarda has a broad host range and geographic distribution, and contains important virulence factors that enhance bacterial survival and pathogenesis in hosts. Although recent progress in edwardsiellosis research has enabled the development of numerous, highly effective vaccine candidates, these efforts have not been translated into a commercialized vaccine. The present review aims to provide an overview of the identification, pathology, diagnosis and virulence factors of E. tarda in fish, and describe recent strategies for developing vaccines against edwardsiellosis. The hope is that this presentation will be useful not only from the standpoint of understanding the pathogenesis of E. tarda, but also from the perspective of facilitating the development of effective vaccines.

Introduction

Edwardsiellosis, caused by Edwardsiella tarda, has been reported worldwide in economically important fish species, including Japanese eel (Anguilla japonica), red sea bream (Pagrus major), yellowtail (Seriola quinqueradiata), channel catfish (Ictalurus punctatus), and turbot (Scophthalmus maximus). This infection also leads to serious economic losses in the aquaculture of olive flounder (Japanese flounder; Paralichthys olivaceus), the most important fish species in South Korean aquaculture, with production valued at 489.7 billion Korean Won (40 922 MT), which corresponds to 56.5% of total fisheries production in 2010.

Recent studies on vaccine development have applied a variety of antigen-preparation methods; however, commercial vaccines are not yet available. In addition, numerous studies have reported on virulence factors of E. tarda and immune responses of hosts. In the present study, the pathogenicities of E. tarda in fish that can be exploited to elicit effective protection strategies against edwardsiellosis will be discussed.

Identification and classification

The genus Edwardsiella is composed of three species, E. tarda, E. ictaluri, and E. hoshinae. Fish are usually infected with E. tarda or E. ictaluri, whereas E. hoshinae infection is usually reported in reptiles and birds. Panangala et al.suggested that biochemical tests can differentiate E. tarda from E. ictaluri among bacteria isolated from freshwater fish based on the positive reaction of E. tarda in tests of indole production, methyl red reduction and hydrogen sulfide generation. In protein profiling of bacterial isolates using sodium dodecyl sulfate-polyacrylamide gel electrophoresis (SDS-PAGE) and Western blotting, the authors also demonstrated that E. ictaluri is more homogenously distributed than E. tarda. E. tarda was originally isolated from cultured Japanese eel (Anguilla japonica) in Japan in 1962.

Subsequent findings for the bacterium were reported from snakes in Japan and from human feces in the USA; the bacterium was designated E. tarda by Ewing et al.. Although there was a move to change the epithet tarda to anguillimortiferum since it had been initially reported as Paracolobactrum anguillimortiferum, the bacterium is commonly named E. tarda because P. anguillimortiferum was not registered and the original culture was lost.

E. tarda is a Gram-negative, short, rod–shaped, facultative anaerobic bacterium that measures about 2–3 ìm in length and 1 ìm in diameter. It is usually motile, but isolates from red sea bream and yellowtail are non-motile. This bacterium can survive at 0–4% sodium chloride, pH 4.0–10.0, and 14–45°C. The biochemical characteristics of E. tarda are catalase positive, cytochrome oxidase negative, production of indole and hydrogen sulfide, fermentation of glucose, and reduction of nitrate to nitrite. However, several variations of biochemical tests have been found for ornithine decarboxylase, citrate utilization, hydrogen sulfide production, and fermentation of mannitol and arabinose. These discrepancies allow division into two groups: wild type and biogroup.

The characteristics of wild-type E. tarda are negative for arabinose, mannitol and sucrose production, and positive for hydrogen sulfide production; the characteristics of biogroup 1 are the opposite.

Park et al.demonstrated that E. tarda can be divided into four serotypes, A, B, C and D, using O-antigen extracts of 445 isolates from infected eel, water, and sediments. They suggested that 72% of isolates belonged to serotype A, the most virulent group based on experimental challenge tests, whereas a subsequent study of E. tarda isolated from olive flounder revealed that all isolates were serotype A. Another study established an E. tarda serotyping scheme comprising 61 O groups and 45 H antigens that is preferable for international applications.

Several important findings suggest that intra- and/or inter-specific variability exists among E. tarda strains. E. tarda isolated from humans could be differentiated from isolates from fish by RAPD (random amplified polymorphic DNA) analysis, and E. tarda isolated from freshwater fish or pond sediments showed diverse and/or homogeneous characteristics in plasmid profiling, ERIC-PCR (enterobacterial repetitive intergenic consensus-polymerase chain reaction), SDS-PAGE, and RFLP (restriction fragment length polymorphism) analyses of 16S rDNA.

In addition, Western blot profiles of LPS (lipopolysaccharides) from E. tarda strains isolated from turbot and other fish revealed that only isolates from turbot were recognized by rabbit sera raised against the isolate from turbot. Biochemical tests, protein profiling, LPS profiling, and RAPD analysis showed that E. tarda strains from olive flounder have highly homogeneous phenotypic and genotypic characteristics compared to isolates from Japanese eel (unpublished data).

Edwardsiella tarda was isolated from 47, 88, and 79% of skin, visceral, and dressed-fish samples, respectively. This species was also isolated from 30% of imported dressed fish, 75% of catfish pond water samples, 64% of catfish pond mud samples, and 100% of frogs, turtles, and crayfish from catfish ponds. The incidence of Edwardsiella increased during the summer months, as water temperatures increased. Of several isolation media evaluated, the most effective was selective enrichment in double-strength Salmonella-Shigella broth and subsequent plating on single-strength Samonella-Shigella agar. The significance of the incidence of Edwardsiella in catfish, catfish disease, and public health could not be substantiated.

Human Infection with Edwardsieila tarda

Edwardsiella tarda, a recently recognized member of the Enterobacteriaceae, has been isolated in the United States and 14 foreign countries and from both cold- and warm-blooded animals. Although the organism has been well characterized bacteriologically, little is known about its role in producing disease in man. A case of liver abscess and septicemia due to E. tarda and the clinical information on eight cases of mild diarrhea or wound infection in which the organism was isolated are reported. In tests on isolates obtained in four cases, the organism was sensitive to antibiotics commonly used in treating infections with gram-negative organisms, including salmonellae. A review of other cases of infection with E. tarda reported to date indicates that this organism is capable of producing disease patterns similar to those caused by salmonellae. The isolation of E. tarda should be regarded as significant if the patient manifests clinical features similar to those described for salmonellosis.

Edwardsiella tarda is a motile, facultatively anaerobic, Gram-negative rod that is categorized as a member of the family Enterobacteriaceae. The genus Edwardsiella was first recognized by Trabulsi et al in 1962, followed by a description of E tarda in the mid-1960s. These organisms have successively been named the "Bartholomew group" by King and Adler, the "Asakusa group" by Sakazaki and "Edwardsiella" by Ewing et al.

E tarda is typically isolated from fresh or brackish water environments such as river mouths. It has also been isolated from the intestines of humans (after eating fresh water food sources such as catfish or eels and from animals, including reptiles and freshwater fish.

E tarda rarely causes infections in humans. The colonization rate in humans ranges from 0.0073% in the Japanese to 1% in Panamanians. Approximately 80% of E tarda infections in humans are characterized as gastroenteritis and E tarda is primarily isolated from stool samples. Extraintestinal infections, such as endocarditis, empyema, hepatobiliary infections, peritonitis, intra-abdominal abscesses, osteomyelitis, wound infections and meningitis, have been reported less frequently. While E tardabacteremia (ETB) occurs relatively rarely (<5%), it can be fatal in humans. However, little is known about the clinical epidemiology of ETB. Therefore, we aimed to document the clinical epidemiology of ETB, including independent risk factors associated with death, the geographical and seasonal distributions, as well as environmental and dietary risks, based on previous reports and recent clinical cases of ETB at our institution (Tokyo Women's Medical Univeristy [TWMU] Hospital [Tokyo, Japan]).

METHODS

A retrospective, observational, case control study was performed. The present analysis comprised literature and retrospective chart reviews. A search was conducted for case reports and case series involving ETB, published between January 1968 and December 2013, with the PubMed MEDLINE and Japanese Medical Abstracts Society ICHUSHI databases using the following keywords: "Edwardsiella tarda", "E tarda" and "infection". An episode of ETB was defined as e"1 positive blood cultures yielding E tarda from the same patient. Results meeting the following criteria were included: published case reports or abstracts from scientific meetings; patients of any age; and articles published in any language including English, French, Spanish and Japanese. Exclusion criteria included: negative blood culture; no description of E tarda infection; animal cases; experimental studies; and review articles.

In addition, the medical records of patients, regardless of age, with ETB admitted to the TWMU (a 1423-bed university hospital in Tokyo, Japan) between April 2000 and December 2013, were included in the retrospective chart review.

The following data were obtained from both published and TWMU-admitted cases that fulfilled the inclusion criteria: age; sex; underlying disease; antibiotic treatment; complications related to the ETB; geographical area; month of ETB onset; history of exposure to environmental risk factors; dietary history to include dietary risk factors; and crude mortality after ETB within 90 days. The case definition for survival required a report of whether the patient survived, or no report of death, in published articles or medical records from cases at TWMU. During the study period, 112,796 blood samples were obtained under sterile conditions and were processed using BACTEC 9240 (Becton Dickinson Diagnostic Instrument Systems, USA) until March 11, 2011, and BacT/ALERT 3D (bioMérieux, France) from April 2011 onward. Strains of E tarda were identified using MicroScan WalkAway 96si (Siemens, Germany). The five cases of ETB identified were included in the present study. The species identification of E tardawas also confirmed using matrix-assisted laser desorption ionization–time of flight mass spectrometry (Microflex LT with MALDI-Biotyper version 3.1, Bruker Daltonik, Germany). Susceptibility of E tardaisolated from blood cultures was tested retrospectively in accordance with the standards of the Clinical and Laboratory Standards Institute using E-test (bioMérieux, France).

Continuous data were compared using a t test and categorical data were compared using Fisher's exact tests; P<0.05 was considered to be statistically significant. Multivariate analyses were used to determine the independent risk factors associated with mortality using forward stepwise logistic regression. All variables with P<0.1 in univariate analyses were entered into the multivariate model.

To our knowledge, the present study was the first and largest multi-language review of ETB. The results indicate that the overall mortality related to ETB is 44.6%, and ETB may be more likely to occur in the humid and subtropical climates of Eastern Asia and near the Gulf of Mexico in the US, particularly during summer and autumn months. Furthermore, environmental and dietary risk factors, such as exposure to brackish water and raw food consumption, may play a role in ETB.

ETB occurs infrequently (<5%) and the risk factors are not well established. Investigating similarities with other species may provide some clues regarding the mechanisms related to ETB. Previous case series suggest that up to 50% of ETB patients also had hepatobiliary diseases, including alcoholic cirrhosis. Interestingly, the authors indicated that conditions resulting in iron overload, such as cirrhosis, sickle cell anemia, leukemia and the neonatal state, are also considered to be risk factors for ETB. In the present analysis, 66.2% of the cases involved hepatobiliary diseases, including cancer. Twenty-one (27.3%) of these cases involved gallbladder stones with recurrent episodes of cholecystitis. The rate of underlying diseases may be higher in ETB than in other E tarda infections; this may indicate that the presence of underlying disease increases the risk for developing ETB. However, the results of our multivariate analyses suggest, for the first time, that liver cirrhosis is an independent risk factor associated with death in ETB. Cholecystitis with gallbladder stones may increase the risk for ETB in a way similar to that of human typhoid fever. Moreover, the overall rate of underlying diseases in ETB may be higher than in other E tarda infections, potentially indicating that the presence of underlying disease increases the risk for ETB and subsequent high mortality from ETB.

In addition, reservoir-related Aeromonas species are similar to E tarda. Both can cause a wide spectrum of diseases among warm- and cold-blooded animals, including fish, reptiles, amphibians, mammals and humans. Population sizes of both species can grow quite large, generally peaking in the warmer temperatures of the summer months in temperate freshwater lakes and chlorinated drinking water, and have been associated with contact wtih reptiles, including those kept as pets. Vibrio vulnificus can also cause severe disseminated infection associated with exposure to seawater and brackish water. Severe V vulnificus infections in humans are also responsible for liver cirrhosis. Therefore, clinical characteristics of E tarda infections in humans resemble those of Aeromonas, typhoid fever and V vulnificus infections, and share similar environmental risks. It is well known that necrotizing fasciitis type 3, which is known as a marine infection and can be fatal within 48 h, is caused by Vibrio or Aeromonas species. Therefore, it should be recognized that E tarda can also result in necrotizing fasciitis type 3. In addition, previous reports indicate that the mortality rate of ETB is nearly 50%, which is similar to that of V vulnificus or severe infections caused by Aeromonas species.

E tarda infections, including ETB, are considered to be foodborne. Despite this, dietary risk factors for ETB have not been well established or documented. In addition, most of the cases of ETB are suspected to occur endogenously through primary colonization and infection in the human intestinal tract. However, documented E tarda infections in humans have resulted from the consumption of infected or contaminated food such as fish. According to the Food and Agriculture Organization of the United Nations Statisitcs Division (FAOSTAT) Food Balance Sheet 2006, Japan has one of the world's highest per capita seafood consumption rates. E tarda has been isolated from raw flounder in Japanese fish farms. Interestingly, E tarda has been reported to result in a 60% to 90% mortality rate among Japanese eels, and a traditional custom in Japan, dating from the 18th century, is to eat cooked eel at the end of July. Furthermore, 71% of the eels from the eel farms in Korea have been reported to carry E tarda.

Toward the end of the 1960s in the US, E tarda was linked to catfish, mainly in the waters of Arkansas, Mississippi, Texas and Louisiana. Despite the small number of cases in the current study, approximately one-half of the cases occurred around the Gulf of Mexico, which receives a supply of nutrient-rich water from several rivers. Individuals residing in these areas may be exposed to high concentrations of E tarda through the water or contaminated food. According to data from the National Fisheries Institute, tilapia, pangasius and catfish are consumed in the US. They are farmed in fresh water and 79% of already-dressed catfish in the US carry E tarda. Given the collective findings of the available literature, we suspect that dietary exposures also affect the risk for ETB. Little is known about the recent prevalence of E tarda colonization in healthy individuals; no recent data are available from large-scale studies in any country. A study conducted in the 1970s, is one of the few to report that only 26 of 353,600 Japanese individuals were healthy carriers of E tarda. We suspect that the rate of colonization in stool appears to be high among individuals in areas with a high consumption of raw fish contaminated with E tarda. The colonization rate in humans may be affected by changes in dietary habits and increased travel, and the use of novel techniques, such as matrix-assisted laser desorption ionization–time of flight mass spectrometry, may improve identification of E tarda, which was previously underestimated.

Trimethoprim/sulfamethoxazole-resistant E tarda has been clinically isolated in a pediatric patient with X-linked chronic granulomatous disease and osteomyelitis in Japan. No multidrug-resistant strains of E tarda were found in the current study. There are several studies showing at least 90% of E tarda strains to be colistin resistant. Natural resistance of E tarda to colistin or polymyxin B has been suggested, but its mechanism was unelucidated..

Feedstuffs for domesticated animals are exposed to colisin in Japan and some European countries. Antibiotic resistance in E tarda may emerge as an issue related to environmental antibiotic exposure in food production. Geographical differences among the Republic of China (Taiwan) and the US in the antimicrobial susceptibility of E tarda, isolated from food-producing animals, have been previously described. Seasonal distribution of human E tarda gastroenteritis has not been previously demonstrated. In addition, information has not been available regarding the time of year for extraintestinal E tarda infections in previously reported cases. The results of the current review suggest that there may, in fact, be a seasonal distribution of ETB, and exposure to higher concentrations of E tarda in the warmer temperatures of the summer months may increase the risk for ETB. For example, Eastern Asia, which includes Japan and the Republic of China (Taiwan), and the states of the US near the Gulf of Mexico, reported the highest rates of ETB and belong to climates that are similar – humid and subtropical.

In addition, dietary customs vary according to geographical region. Therefore, dietary factors may play a role in the geographical or seasonal distributions of ETB.

Taking all of this into account, we propose two reasons for the high prevalence of ETB in Japan. The first is that Japan is geographically rich in brackish water areas and rivers. Second, the Japanese consume a relatively large amount of raw seafood, which may be contaminated with E tarda.

The present review had limitations. One was the retrospective nature of the data, including the use of previously reported literature and only five recent cases from our institute. Fourteen of 72 cases from the literature were described only in the Japanese language. Therefore, publication and resource bias may have affected our results. The second limitation is that we focused on the results of blood cultures; however, sensitivity of blood cultures may vary during a long study period.

ETB may be categorized as a severe food- and waterborne infection similar to Aeromonas, Vibrio and Salmonella (typhoid fever) infections, which result in high mortality in patients with severe underlying diseases such as liver cirrhosis. Geographical and seasonal distributions may characterize ETB. The clinical epidemiology, including dietary risk factors and the current incidence of ETB should be further established.

Myonecrosis Caused by Edwardsiella tarda: A Case Report and Case Series of Extraintestinal E. tardaInfections

Human infections caused by the bacterium Edwardsiella tarda are rare. E. tardahas been cultured from samples of human feces, blood, urine, CSF, bile, peritoneal fluid, and wounds. In >80% of cases, the organism is cultured from stool specimens and is associated with gastrointestinal illness. The precise role of E. tarda as a pathogen has been obscured by its isolation from fecal samples obtained from persons without gastrointestinal symptoms and by its frequent isolation with other known pathogens. We report the first case to our knowledge of myonecrosis due to infection with E. tarda in a presumed immunocompetent patient. We also review the experience with E. tarda at the Medical Center of Louisiana in New Orleans during the 7 years from 1993 through 1999.

Case Report and Methods

A healthy 48-year-old man fell into brackish water while crab fishing and lacerated his forearm by contact with a submerged brick. His wound was irrigated and sutured in an emergency department. He received no antibiotics. The next day the patient developed pain, swelling, and a purulent drainage from his wound. He returned to the hospital and was treated with iv piperacillin/tazobactam, penicillin, and doxycycline. He underwent a fasciotomy and surgical debridement of necrotic fascia and muscle. Cultures of samples obtained during surgery yielded 1 organism, E. tarda. Results of blood cultures were negative. This isolate was sensitive to all antibiotics tested, including ampicillin, ampicillin/sulbactam, ticarcillin, gentamicin, cefazolin, cephalothin, trimethoprim-sulfamethoxazole, piperacillin, imipenem, and ticarcillin/clavulanate. The patient had an uncomplicated postoperative course. This study was a chart review of all isolates of E. tarda cultured from clinical specimens at the Medical Center of Louisiana in New Orleans from 1 January 1993 through 31 December 1999.

RESULTS

Eleven cases were identified. All infected patients survived. Five cases were associated with wound infections, and 3 of the wounds occurred in a marine environment: an arm laceration that resulted from a fall into brackish water (patient 1), a leg laceration that resulted from a fall into a canal (patient 2), and a puncture wound to a foot that happened after the patient stepped on fish bones (patient 3). Patient 4 had a hand laceration caused by broken glass. Patient 5 had infection of a finger after it was accidentally crushed in a car door. All 5 patients required surgical incision and drainage. In 2 of these patients (patients 1 and 4), E. tarda was isolated in pure culture. Cultures of wound samples obtained from patients 2 and 5 yielded E. tarda and Aeromonas hydrophila, and cultures of samples from the wound of patient 3 yielded E. tardaand Shewanella putrefaciens.

Two women developed gynecologic infections that involved abscess formation. In the first woman (patient 6), E. tarda was isolated in pure culture from a ruptured tuboovarian abscess. The second woman (patient 7) had a Bartholinian abscess; on culture, samples of the abscess yielded isolates of E. tarda, Escherichia coli, and ã-hemolytic streptococci.

Two additional cases in which abscesses were described included a thumb felon (in patient 8), which yielded E. tarda, coagulase-negative staphylococcus, and Fusobacterium species on culture, and a perirectal abscess (in patient 9), which yielded E. tarda and group C streptococcus on culture.

E. tarda and Klebsiella pneumoniae were isolated from a patient with necrotic cholecystitis (patient 10). The final patient (patient 11) was a previously healthy man who presented with diarrhea, melena, and abdominal pain. He received a diagnosis of upper gastrointestinal hemorrhage secondary to peptic ulcer disease. Cultures of blood samples yielded E. tarda.

Susceptibility testing, done by use of semiautomated broth microdilution method (VITEK; bioMériux Vitek), revealed that all of the E. tarda isolates were sensitive to the following antibiotics: ampicillin, ampicillin/sulbactam, ticarcillin, gentamicin, ofloxacin, cefazolin, cephalothin, piperacillin, ticarcillin/clavulanate, and imipenem.

E. tarda, a member of the family Enterobacteriaceae, exists widely in nature. It is associated with freshwater environments and it has been cultured primarily from the samples of stool of animals that inhabit these ecosystems, including fish, lizards, toads, snakes, turtles, crayfish, and alligators. It has also been cultured directly from samples from lakes, rivers, wells, catfish ponds, and from sewage water. Although E. tarda has not been directly cultured from seawater, it has been cultured from autopsy specimens of ocean inhabitants, such as sea lions and penguins, Almost all isolates of E. tardaare uniformly susceptible to antibiotics with activity against gram-negative bacteria (i.e., aminoglycosides, cephalosporins, â-lactams, and fluoroquinolones).

The most frequently reported manifestation of infection caused by E. tarda in humans is gastrointestinal disease. The organism has been cultured from stool samples obtained from patients with diarrheal illness in Asia, Australia, and Central America. One early report described 4 different clinical presentations: mild gastroenteritis, a typhoid fever-like illness, bacteremia, and an asymptomatic or carrier state. Later reports confirmed the isolation of E. tarda from cultures of stool samples obtained from patients with gastrointestinal illness as well as those without symptom. The role of E. tarda in acute gastrointestinal illness in some patients may be confused by the concurrent isolation of other known gastrointestinal pathogens, such as Salmonella and Shigella species.

Extraintestinal infection with E. tarda is considered unusual. E. tarda-associated sepsis, although uncommon, has a mortality rate that approaches 50%. In one review, 11 of 14 cases of E. tarda sepsis occurred in patients with significant underlying medical conditions, particularly hepatobiliary disease or hemoglobinopathies. E. tarda has also been implicated in a variety of infections, including reports of isolates from patients with osteomyelitis, peritonitis, cholecystitis, and salpingitis.

Wound infections with this organism often involve abscess formation following trauma. Several cases have occurred in previously healthy people who were exposed to E. tarda in an aquatic environment, such as a patient with an infected foot laceration acquired while walking on the shore of a lake and a patient with an infected puncture wound caused by a catfish. In both cases, E. tardawas isolated with A. hydrophila, an organism known to cause infection in wounds that are exposed to fresh water.

The only patient with a necrotic deep soft-tissue infection caused by E. tardawas a 67-year-old man with cirrhosis and hepatocellular cancer who was admitted to the hospital with massive ascites and edema of his lower extremities. On day 15 of hospitalization, he became febrile and developed diarrhea. Cellulitis with hemorrhagic bullae of the legs followed, and he died of septic shock. Histopathological examination revealed necrotizing fasciitis. Cultures of samples of blood and fluid obtained from the bullae yielded E. tarda. In the present series, only extraintestinal sites of infection with E. tarda were identified. The microbiology laboratory at the Medical Center of Louisiana in New Orleans does not routinely isolate and identify all gram-negative bacilli recovered from cultures of stool samples. Salmonella, Shigella, Yersinia, Campylobacter, Vibrio, Aeromonas, and Plesiomonas species and E. coli O157 are identified and reported. No isolates of E. tarda were recovered from stool samples. However, it is likely that this practice results in underreporting. In fact, 2 of the patients we studied who had E. tarda in pure culture, 1 patient with tuboovarian abscess (patient 6), and 1 patient with sepsis (patient 11) experienced bloody diarrhea. It is possible that E. tarda could have been isolated from samples of their stool had it been sought.

Our case series includes the first report of E. tarda as a cause of myonecrosis in an immunocompetent patient. Of interest, 3 of the 5 cases in which E. tarda was isolated from wound infections (in patients 1–3) involved exposure to a marine environment, a likely source of the organism. In addition, 3 patients with wound infections (patients 2, 3, and 5) had cultures that yielded other known waterborne pathogens (i.e., A. hydrophila and S. putrefaciens). Patients 1 and 2 had wounds that were exposed to brackish water, and their cultures yielded E. tarda alone and E. tarda with A. hydrophila, respectively. Although it is not common, serious soft-tissue infections that follow exposure to fresh- or saltwater environments, such as those due to Vibrio vulnificus, are well described, and some authors recommend prompt surgical debridement and antibiotic therapy to reduce the risk of posttraumatic wound infection.

Seven of the 11 cases in this series, including the 3 that involved marine exposure, were identified during the warmer weather months of May through October. This association may reflect an increase in recreational aquatic exposure during warmer weather, increased proliferation of E. tarda in warmer water, or both. In a study reported elsewhere, cultures of samples of mud and water from ponds in Texas yielded E. tarda, and the quantity of this organism increased during the summer months. No seasonal distribution of E. tardagastroenteritis has been demonstrated, and no information regarding the time of year of previously reported cases of extraintestinal E. tarda infection is available.

Two patients in this series presented with gynecologic infections (patients 6 and 7). Only 3 other cases of gynecologic infection associated with E. tarda have been reported. The first occurred in a young woman with salpingitis, for whom no further information was provided. The second involved a healthy 42-year-old native Hawaiian woman with a tuboovarian abscess ; the authors speculated that the infection was related to colonization of the gastrointestinal tract after ingestion of raw fish. The third patient described was a 46-year-old woman who was admitted to the hospital with fever, watery diarrhea, abdominal pain, and vaginal spotting. Culture of blood yielded E. tarda. She received treatment with iv antibiotics and eventually underwent hysterectomy; histopathology revealed a uterine myoma with infarct and abscess formation. This patient had a history of hepatobiliary disease, and she had a habit of eating raw meat, suggesting a possible association of infection with the gastrointestinal system. The patient in our series with a tuboovarian abscess (patient 6) also experienced fever, abdominal pain, and diarrhea. However, she had no exposure to freshwater or saltwater environments, she rarely ate seafood, she always cooked her food, and she had not recently handled any wild animals or reptiles. The portal of entry for infection remains unknown.

Four patients in our series had E. tarda isolated in pure culture (patients 1, 4, 6, and 11), which establishes the capability of this organism to cause limb- and life-threatening infection. Three wound infections (in patients 2, 3, and 5) were polymicrobial and included other pathogens that are known to cause waterborne wound infections (A. hydrophila and S. putrefaciens). In the 4 remaining cases of E. tarda infections (in patients 7–10), other bacterial pathogens were also isolated. The precise role of E. tarda in the cases of mixed infection is not clear. Because this case series also demonstrates the ability of E. tarda alone to cause serious extraintestinal infection, it is likely that the organism is a copathogen in the mixed infections rather than a commensal organism.

None of the 11 patients in our series had significant underlying medical conditions. Although 1 patient had sickle cell trait (patient 2) and 1 tested positive for hepatitis C antibody (patient 6), both were clinically well and neither exhibited signs or symptoms of their diseases. The antibiotics administered for treatment in this series varied.

In conclusion, human infection with E. tarda is uncommon and manifests a diverse clinical spectrum ranging from colonization of the gastrointestinal tract to life-threatening necrotizing soft-tissue infections and sepsis. Eleven patients with extraintestinal infection are reported, including 1 patient with sepsis and 10 patients with abscess formation that required surgical drainage. Seven of the 11 cases occurred during the warmer weather months of May through October. Underlying immunosuppressive conditions (i.e., hematologic or hepatic disease) have previously been reported as risk factors for infection, although none of the patients in this series had chronic illnesses. Exposure to marine environments appears to be a risk factor for wound infections. E. tardacan be isolated with other pathogenic bacteria in culture, which can create some confusion regarding its role as a pathogen. However, increasing numbers of reports (including 4 patients in this series) describe serious infections secondary to E. tarda alone, which provides evidence that this organism can cause significant morbidity and death. All isolates of E. tarda were susceptible to commonly used antimicrobial agents directed against gram-negative organisms, which is consistent with previous reports. This case series contributes to a small but growing body of literature that establishes E. tarda as an unusual but potentially life-threatening pathogen.

Edwardsiella tarda was the first species identified of the genus Edwardsiella, and was named after a renowned microbiologist P. R. Edwards (Janda, 1991). E. tarda was originally named Edwardsiella anguilimortifera, but it was ultimately changed to E. tarda because this name was used more often in scientific reports. E. tarda is a Gram-negative bacilli that belongs to the Enterobacteriaceae family and was first characterized in 1965 (Health, 2001). E. tarda has many traits that are characteristic of many enterobacteria such as E. coli. These characteristics include it being a facultative anaerobe, rod-shaped, and motile (Health 2001). Its motility is due to peritrichous flagella. Although Edwardsiella tarda was initially characterized more than thirty years ago, there is still very little known about this bacterium. E. tarda is known for causing diseases in both humans and fish, both of which can potentially be fatal if untreated. Though this may be the case, the likelihood of a serious infection is very slim. As a fish pathogen, it is of particular importance to aquaculture and the fishing industry, especially commercial fish farms. It may become more of a significant health issue to fish and humans alike, especially in light of emerging and increasing antibiotic resistance in fish pathogens, due in large part to overuse of antibiotics in fish farming (Greenlees et al., 1998; Lehane and Rawlin, 2000). Some studies have focused on using proteomics and molecular techniques to elucidate the mechanism of pathogenesis in Edwardsiella tarda (Rao et al., 2004). Studies such as these have allowed the characterization of novel toxin secretion pathways, such as the discovery of a type VI secretion system essential for E. tarda pathogenesis (Zheng and Leung, 2007). These types of analyses help us better understand bacterial pathogenesis in general, as well as provide new insights for fighting disease.

Edwardsiella tarda is predominantly found in freshwater environments colonizing the guts of fish, including flounder living off the coasts of East Asia and catfish that live in regions of the United States. It can also be found in the intestinal tract of birds, reptiles, and mammals (Janda et. al, 1991). It can be found as part of the human intestinal flora as well, although this seems to be rare, and is usually due to an infection or prolonged contact with contaminated water. Other factors that increase the risk of getting an infection include high iron concentration, very young or old age, or an immune incompetence.

Recent research has suggested the increasing presence of E. tarda in Antarctic wildlife including those animals that live outside aquatic systems. According to Leotta and colleagues, two reported cases of injured or wounded wildlife with E. tarda infections led to performing a study of the prevalence of E. tarda in the Antarctic wildlife. In their study, "The classic Edwardsiella tarda was isolated from 281 (15.1%) of 1,855 Antarctic wildlife samples obtained from southern giant petrels, brown skuas, greater sheathbills, Adelie penguins, gentoo penguins and chinstrap penguins from Potter Peninsula in 2000 and 2002. " (Leotta et al., 2009). Samples of feces, eggs, and other means were taken from those various species. "[They] considered that E. tarda was a common bacterium in the feces of Antarctic birds and mammals, because no significant differences were found with regard to the areas and seasons investigated. This was supported by the fact that there were no significant differences in the distribution of positive rates obtained within the species evaluated. However, E. tarda was described as an opportunistic bacterial pathogen and has been commonly isolated from different healthy, sick and moribund Wsh, birds and reptiles in several ecosystems around the world." (Leotta et al., 2009). The study concluded that the presence of E. tarda was common among many species in the Antarctic, although more tests had to be run to see other possible pathogenic traits. The presence of E. tarda even in smaller numbers as reported by this study only highlights the possibility of the spread of disease in Antarctic wildlife and further study may help with identifying and evaluating the condition of this and other ecosystems.

PATHOLOGY

E. tarda is typically found in the normal gut flora of fish and humans, and can be an opportunistic pathogen in human, causing gastroenteritis and diarrhea (Verjan, 2005). E. tarda has a high affinity for red blood cells due to specific fimbriae that it produces. As a result, Edwardsiella tarda has hemagglutination properties (Sakai, 2003). The fimbriae are encoded in the genome as a region made of 534 base pairs (Sakai, 2003). Scientists have found that the hemagglutination properties of E. tarda is inhibited by N-acetylneuraminic acid and fetuin, but not D-mannose (Sakai, 2003). In the lab, E. tarda is easily grow on nutrient agar plates and nutrient broth. Though Edwardsiella tarda was first characterized in the 60's, there have been few studies done to test its relative pathogenicity. However, the few research that has been conducted on it shows that it releases dermatotoxins that damage the skin. Studies also show that E. tarda is invasive in HEp-2 cell monolayers, and that it produces a cell-associated hemolysin and siderophores (Janda et al, 1991). Siderophores are iron-binding molecules produced by freshwater and marine bacteria. These siderophores are related to microbial virulence, and scavenge the iron inside of host cells, providing the microbial cell essential micronutrients (Janda et al, 1991). E. tarda flagellar genes, fliC12, fliA and flhDC are essential for the number and length of flagellar filaments, and subsequently the swimming and swarming ability of E. tarda. These genes have also been shown to aid biofilm formation ability, adherence and internalization to Epithelioma papulosum cyprini (EPC) cells, and pathogenicity to zebrafish. E. tarda has been shown to be resistant to the bactericidal properties of host serum by preventing serum complement activation via the alternative pathway. This important virulence factor is facilitated by several host serum-induced proteins, including the novel putative zinc metalloprotease (Sip1). When used as a subunit vaccine, purified recombinant Sip1 (rSip1) effectively induced protection in flounder fish against E. tarda infection. The type III secretion system (T3SS) of E. tarda enables infection by translocating effector proteins into host cells. The E. tarda T3SS gene cluster consists of 34 genes, including the esrA-esrB genes which contribute to the virulence of E. tarda. The gene esrB is responsible for the secretion of translocon proteins like EseB. EseB mediates autoaggregation and subsequently biofilm formation of E. tarda by forming filamentous appendages on the bacterial surface.As there is iron deprivation in the host, the CAH (cell-associated hemolysin) of E. tarda provides iron by cleaving erythrocytes and releasing hemoglobin. Because E. tarda is so prevalent in the freshwater environment, it is easy to see why many of the organisms living within these areas will carry E. tarda within their intestinal tract. Many gram-negative bacteria (including E. tarda) use ferric uptake regulator (Fur) as an important transcriptional regulator.

In E. tarda, Fur's role is significant and multifaceted; Fur affects the growth, siderophore production, and acid tolerance of E. tarda, helps protect against oxidative stress and host serum, helps inhibit host immune response, and generally increases the overall virulence of the bacterium. This microorganism is an opportunistic pathogen, and is able to live inside and outside of the host organism. One of the most common diseases among freshwater fish is hemorrhagic septicemia, also known as edwardsiellosis, which usually results in the death of the fish (Yousuf, 2006). This disease is usually prevalent in flounder. Pathogenic E. tarda was first isolated and documented in farmed striped catfish (Pangasianodon hypophthalmus) in fish sampled between October 2007 and May 2008 from a farm in the Bhimavaram district of Andhra Pradesh, located on the east coast of India. A higher prevalence of infection was observed in the summer as opposed to the winter. A few known epidemics of edwardsiellosis have been recorded, and can be devastating to the fish population, but if the infection is detected early, the epidemic may be easily avoided (Yousuf, 2006). Two outbreaks of Edwardsiellosis were recorded in October 2013 and July 2014 in cage-cultured sharpsnout sea breams, Diplodus puntazzo, in a single commercial fish farm located in the Saronikos Bay of Eastern Greece. This was not only the first report of Edwardsiellosis in this species, but also the first report of E. tarda in cage-cultured fish in the Mediterranean, in general. In humans, E. tarda can cause gastroenteritis, which is infection in the stomach and intestines (Clarridge, 1980). This is especially true in areas where raw fish is a large part of their daily diet. E. tarda can also cause colitis and dysentery-like diseases in humans (Janda et al., 1991) as well as gastroenteritis, infections of wounds, gas gangrene associated with trauma to mucosal surfaces, and systemic disease such as septicemia and meningitis (Janda and Abbott, 1993). These gastroenteritis cases are usually misdiagnosed and attributed to other agents. The best way to avoid an E. tarda infection would is to not spend too much time in contaminated water, and practice good hygiene habits.

E. tarda can even be transferred to infants during the birthing process. Mowbray and her colleagues described a case where a 6-day old boy had E. tarda growing within its gastrointestinal tract since the time of birth. The strain of E. tarda was the same as the one found colonizing the mother's vaginal and gastrointestinal areas, as demonstrated by both fingerprinting and antibiotic susceptibility analyses (Mowbray, 2002). The mother had spent time in contaminated lake water during her pregnancy, demonstrating that exposure of the mother to contaminated water poses a risk to newborn babies (Mowbray, 2002). The incidence of E. tarda infection in newborns is increasing. Before 2003, there were only two known cases of neonatal sepsis caused by Edwardsiella tarda (Mowbray, 2002). By 2003, however, there were a total of 300 cases of E. tarda infection, in which 83% of the patients had gastroenteritis. Studies showed that E. tarda is able to become part of the vaginal flora and then cause neonatal sepsis through infection at birth. Very little is known about the vaginal colonization of E. tarda, but there were only a handful of cases where patients had gynecologic infections (Mowbray, 2002). E. tarda has hemolytic properties which can cause reddening and systemic hemorrhagic septicemia in infected fish. Hemolysis is facilitated by the pore-forming cytotoxic protein, Cytolysin A (ClyA), which is encoded by the clyA gene. Eha protein was found to act as an important transcriptional regulator of clyA, and thus also of general hemolytic activity in E. tarda.

Although Edwardsiella tarda may cause infection in humans – some of which can become lethal, there are many ways to prevent infection. E. tarda is still very susceptible to many antibiotics, including ampicillin, antifolates, chloramphenicol, ciprofloxacin, kanamycin, most â-lactams, and nitrofurantoin (Stock, 2001). On the other hand, these bacteria are very resistant to colistin, glycopeptides, lincosamides, streptogramins, and rifampin (Stock, 2001). Compared to the other species in the genus Edwardsiella, E. tarda is also resistant to oxacillin and benzylpenicillin (Stock, 2001).

Other types of treatment that would inhibit E. tarda's growth include moist and dry heat, usually above 121ÚC or above 160ÚC respectively. Edwardsiella tarda is also susceptible to many disinfectants such as 70% ethanol, iodine, and formaldehyde (Stock, 2001).

In their article, "Human Edwardsiellosis Traced to Ornamental Fish," Vandepitte et. Al review the pathogenic role of Edwardsiella tarda and the role of that "ornamental" fish play in transmitting infections. E. tarda can cause human Edwardsiellosis, an intestinal infection. It is most prevalent in tropical areas such as Singapore, Vietnam, and Cuba. The article reviewed a case report involving a two month old infant male in suburban Belgium, and had two days of vomiting. The infant did not have a fever and the stool cultures were normal. Following this initial exam it was the reported that the infant experienced diarrhea, and unformed stools that had a yellowish color and contained mucus. This time the stool culture came back positive for E. tarda. This was the first documented case of a human Edwardsiellosis in Belgium. The patient was then followed for two months, in which the diarrhea continued and additional stool cultures remained positive for E. tarda. Antibiotics where prescribed for two weeks, after which the stool cultures were normal. The patient's home was found to have an aquarium which contained ornamental fish. A Pterophyllum scalarem, a tropical fish that originates in Brazil, was retrieved. The fish was was cultured and tested positive for Citrobacter freundii, Aeromonas hydrophila, and E. tarda. The authors write, "The introduction of ornamental fish into pediatric hospitals presents a potential hazard for neonates and young infants" (Vandepitte 166) because they are potential vehicles for transmission. This case study brings to light that contact with tropical fish, even in an aquarium setting, may increase your risk of transmitting and/or becoming infected by bacterial pathogens such as E. tarda.

Diagnosis

Edwardsiellosis in fish usually occurs under imbalanced environmental conditions, such as high water temperature, poor water quality, and high organic content. Fish infected with E. tarda show abnormal swimming behavior, including spiral movement and floating near the water surface.

Although clinical signs vary after onset, fish infected with E. tarda show loss of pigmentation, exophthalmia, opacity of the eyes, swelling of the abdominal surface, petechial hemorrhage in fin and skin, and rectal hernia. Internally, watery and bloody ascites in the abdominal space and congested liver, spleen, and kidney are found.

Histopathological characteristics of edwardsiellosis in fish are suppurative interstitial nephritis, suppurative hepatitis, and purulent inflammation in the spleen. Abscesses of various sizes, bacterial colonization, and infiltration of neutrophils and macrophages are found in the liver, spleen, and kidney. Some remarkable pathological features have also been demonstrated in fish, such as dorsolateral petechial hemorrhage and abscesses in cutaneous lesions of channel catfish; hyperplasia, necrosis and inflammation in lateral line canals of striped bass; and necrosis and aggregation of bacteria-laden macrophages in red sea bream. However, the symptoms and pathological changes in fish are similar to those of other bacterial infections, including Aeromonas hydrophila, Vibrio anguillarum and Pseudomonas anguilliseptica; thus, other molecular or biochemical methods are recommended for diagnosis of E. tarda infection.

Effect of common aquaculture chemicals against Edwardsiella ictaluri and E. tarda

Edwardsiellosis is an important bacterial infection of freshwater and marine fishes. Edwardsiella ictaluri causes enteric septicemia of catfish, and E. tarda causes emphysematous putrefactive disease of catfish and fish gangrene in various species; these diseases have considerable economic effects on the aquaculture industry. In addition, E. tarda is an important zoonotic pathogen. Thus, the reduction or elimination of these pathogens from an aquarium or aquaculture facility is imperative. This study examined a variety of commercially available chemicals for their ability to reduce or eliminate E. ictaluri and E. tarda from the aquatic environment. The various concentrations of chemicals were tested in vitro in microcentrifuge tubes with a known concentration of bacteria at room temperature. In this study, ethyl alcohol (30, 50, or 70%), benzyl-4-chlorophenol/phenylphenol (1%), sodium hypochlorite (50, 100, 200, or 50,000 mg/L), n-alkyl dimethyl benzyl ammonium chloride (1:256), povidone iodine (50 or 100 mg/L), glutaraldehyde (2%), and potassium peroxymonosulfate/sodium chloride (1%) were effective disinfectants, as each reduced or eliminated the number of detectable organisms within 1 min of contact time. However, neither Chloramine-T (15 mg/L) nor formalin (250 mg/L) substantially reduced bacterial counts even after 60 min of contact time.

A study was carried out from October, 2009 to April, 2010 with the objective of isolating Edwardsiella tarda an important fish pathogen from fish harvested for human consumption from Lake Zeway and Langanoo. A total of 372 tissue samples (three from each fish) comprising liver, intestine and kidney were collected from 124 fish (Clarias gariepinus and Oreochromis niloticus originated from Lake Langanoo and Zeway. Distribution of E. tarda infection among the three organs examined indicated that E. tarda was isolated most frequently from liver (6.5%) followed by intestine (2.4%) and kidney (0.8%) with significant difference among organs. Statistical significant differences (P<0.05) were found in E. tarda infection with respect to site although the bacterium was isolated from fish originating from both Lake Zeway and Langanoo with E. tarda being more prevalent in fish sampled from lake Zeway. E. tarda was isolated more frequently from male fish, the differences in the occurrence of E. tarda infection with respect to sex were not significant (P>005) indicating that both sexes are equally susceptible. The isolation of Edwarsiella from wild fish population of Lakes Zeway and Langano destined for human consumption in the current study is indicates that E. tarda is a potential threat of both the fishery sector/aquaculture and public health. Finally, as is the case for any infectious fish pathogen, there is limited information on E. tarda of fish in Ethiopia and hence further study to have comprehensive information on the agent is forwarded.

Aquaculture is growing rapidly worldwide with fish being the primary sources of animal protein in many countries. The fishery sector plays a significant role in food security through supplementation of food for developing countries. As a whole fish currently make up about 19% of the total protein consumption or just over the 5% of proteins from both plants and animals origin.

As in all animal production systems, however, fish are possibly susceptible to microbial diseases which are one of the major problems hampering production, development and expansion of the aquaculture industry. Fish diseases are global problem affecting fresh water, marine water, cultured, sport and even ornamental fish. The problem is extremely important when fish are subjected to intensive culture practices.

The control of fish diseases is particularly difficult because fish are often farmed in system where production is dependent on natural environmental conditions. Changes or deterioration in the aquatic environment cause the occurrence of most fish disease and also environmental effects play a great role in influencing the health status of fish. Therefore, the multidisciplinary approaches involving the characteristics of potential pathogenic microorganisms for fish, aspects of the biology of fish as well as a better understanding of the environmental factors affecting such cultures will allow the application of adequate measures to prevent and control the diseases limiting fish production.

Edwardsiellosis is among the most important bacterial diseases causing severe economic losses in fish farms of many countries. The disease is caused by E.tarda which is a gram negative, motile, facultative anaerobic, short rod shaped bacterium (1ìm in diameter and 2-3 ìm long) pathogenic to a wide range of fish hosts such as Channel cat fish (Ictaluri punctatus), Striped bass (Morone saxatili), eel (Anguilla anguilla), Tilapia (Oreochromis niloticus), carp (Cyprinus cyrpio) and Flounder (Paralichthys olivaceus . The organisms is frequently found in organically polluted water, poor quality water and affect fish stressed by this situations. E.tarda can be isolated on Edwardsiella Isolation Media (EIM), Brain Heart Infusion (BHI), Tryptic Soya Agar (TSA), Xylose Lysine Deoxycholate(XLD) and MacConkey. It is seen as small, circular, raised, whitish with black center on XLD and pale on MacConkey agar, grow best at a temperature between 25°C-37°C, PH 7-8 and 0.5%NaCl, and characteristically, catalase positive, cytochrome oxidase negative, glucose fermentative, indole positive, citrate negative, lysine positive, mannitol, dulcitol, sorbitol, inositol, xylose, rhamnose negative, produce hydrogen sulfide, alkaline slant and acid but on Triple sugar iron Agar.

Edwardsiella tarda is a health threat not only to fish and other animals but also to humans with the risk factors being exposure to aquatic environment, pre-existing liver diseases, iron over load and raw sea food ingestion.

In humans the bacteria usually cause diarrhea, gastroenteritis, wound infection and even death. There are reports of extra intestinal infection with the clinical pictures including a typhoid likc illncss, peritonitis with sepsis and cellulites with occasionally liver absecess and meningitis. The infection is more severe in immunocompromised individuals.

The practice of consuming partially cooked fish meals, manual handling of fish and unhygienic practice during filleting in Ethiopia indicate that the public is at higher risk of contracting the disease. Therefore, the disease deserves attention due to its impact on the fishery sector as well as its potential threat to future aquaculture industry and public health. In Ethiopia, the bacterium has been isolated from apparently healthy fish of Lake Zeway and Tana. However, there is no further work done in covering the different fish species and aquatic environments. Therefore, this study was conducted with the aim of Isolating E. tarda from cat fish and Tilapia slauthered at zeway fishery resource center originating from Lake Zeway and Langanoo and elucidates the safety of fish products with respect to Edwardsiella tarda contamination.

MATERIALS AND METHODS

Study site

The study site comprised fish species harvested from lakes Zeway and Langanoo.

Lake zeway

Lake zeway is located on the Eastern side of Zeway town, 163 km South east of Addis Ababa it lies in northern part of the rift valley between 7°51N to 8°7'N and 38'43' E 38°57' E with an open water area of 422 km and shoreline length of 137 km. The lake is fed by two major rivers, i.e. Ketar and Meki River and has one out flow in the South, Bulbula river which flow into Abiyata. Five bigger islands are situated in the lake Viz Tulu Gudo (4.8 km), Tsedecha (2.1 km), Funduro (0.4 km), Debresina (0.3 km) and Galila (0.2 km). While the latter two have few inhabitants, the three bigger ones are populated with several hundreds of people.

The catch from Lake Zeway consists of almost exclusively Tilapia (Oeochromis niloticus). Since recent years, however, Cat fish (Carias gariepinus) and Crucian carp (Carcasius gracius) have appeared in small amounts of the total catch. There are a number of landing points around the lake from where fish is collected either by boat or trucks and brought to the major landing points adjoining Zeway town.

Lake langanoo

Lake langanoo is located 200 km South of Addis Ababa lying between 7°36' N;38°45' E. It is 18 km long and 16 km wide with an open water area of 230 km , 7.5 km shoreline and 1600 km catchments area. The main fish species in the lake include Barbus species, Clarias species and Oreochronis niloticus with the total annual catch of 1000 tones.

Study animals

Study animals used in this study included African cat fish (Clarias gariepinus , N=30) and Nile tilapia (Oreochromis niloticus ; N=94) which were harvested from Lake Zeway and Langanoo for human consumption. The fish were physically examined for any external lesions before necropsy and collecting tissue samples.

Necropsy and Tissue Sampling

In dissecting fish, ventral approach to kidney was employed. The fish sample was cut along the midline of the abdomen starting from the anus up to the mouth using sterile dissecting scissor followed by another dissection from the anus to the lateral line and further along the lateral line up to the gills cover to remove the lateral side of the abdominal wall and expose the internal organs. Internal organs were examined for any gross pathology and the findings recorded. Tissue samples were then taken from intestine (N=124), liver (N=124) and kidney (N=124) aseptically using sterile scalpel blade and forceps kept in sterile universal bottles of 100 ml capacities. All necropsy and tissue sampling procedures were carried out under asceptic condition. The bottles containing the samples were then kept in ice box containing ice packs all the way to School of Veterinary Medicine, Debre Zeit where they are processed for bacterial isolation and identification.

Laboratory Examination

Isolation of Edwardsiella tarda

Tissue samples from kidney, liver and intestine of cat fish and tilapia were homogenized in physiological saline. The homogenate was then taken by sterile loop and streaked on xylose lysine deoxycholate agar plate (Titan Biotech) and then incubated at 37°C for 24 hours. Colonies showing or resembling with morphological characteristics of E.tarda were further subcultured on MacConkey agar plates and incubated at 37°C for 24 hours. All lactose non-fermented colonies were further subcultured on tryptic soya agar containing 0.5% NaCl and incubated at 37°C for 24 hours. Presumptive identification of the resulting isolates (colonies) was done employing different tests which included primary bacteria identification techniques and biochemical identification tests.

Primary identification of isolates

Primary Identification of pure culture of the isolates was done based on gram reaction, motility tests, and catalase and oxidase tests. Overnight cultures of pure colony on tryptic soya agar (TSA) plates were used in all of these tests.

Gram reaction

Gram staining was done according to the procedure described by Rowland et al. accordingly, colonies that were gram negative, short rods were considered for further tests.

Catalase test

Catalase test detects whether the bacterium has the enzyme catalase that converts hydrogen peroxide to water and gaseous oxygen. The test was carried out on pure fresh colony on tryptic soya agar plates. Since Edwardsiella tarda is catalase positive, colony showing an elaborated bubble formation was considered positive and taken for further tests.

Oxidase test

This test detects the presence of cytochrome oxidase enzyme in a bacterial cell and characterized by purple colour formation within 10 seconds when the bacterial sample is made in contact with 1 percent aqueous solution of tetramethyl-pphenylenediamine dihydrochloride. In this study, oxidase test was conducted employing filter paper method for each bacterial isolates.

Motility test

Motility test was conducted using sulfur, indole and motility test media. The slant of the medium in test tubes was stabinoculated with fresh colony (isolate) using a straight sterile wire followed by incubation at 37°C for 24 hours. Turbidity of the medium or outgrowths from the line of inoculation was considered indicative of motility and the results recorded.

Biochemical tests

Secondary biochemical identification of bacterial isolates was conducted employing conventional biochemical tests according to the standard procedures described previously. These tests are based on the ability of the bacterium to utilize a sugar, an amino acid or an alcohol or any carbon source in the medium where by the byproducts of such reaction if any is detected using appropriate indicators incorporated into the medium. The detailed procedures of the biochemical tests are presented. In all of the tests, care was taken to maintain asceptic procedures to avoid contamination.

Triple sugar iron (TSI) test: Triple sugar iron agar (TSI) test shows hydrogen sulfide production, gas production, fermentation of lactose, sucrose and glucose. E. tarda is expected to show red slant and yellow butt with hydrogen sulfide production. In this work, TSI test was carried out by inoculating (by stabbing the butt and streaking the slant) of the test tube of TSI agar slant using straight inoculating wire after which the inoculated tube was loosely capped and the findings recorded after 24 hours of incubation at 37°C.

Indole production, motility and H2S production tests

SIM media (BBL) was used to demonstrate indole production, motility and hydrogen sulfide (H 2S) production. SIM media in test tubes were inoculated with pure overnight grown colonies on TSA plates followed by incubation at 37°C for 24 hours after which the findings were recorded. To demonstrate if there is any indole production, kovac's reagent was added to SIM media after 24 hours of incubation and deep red colour was developed.

Simmon's citrate test

Simmon's citrate slants in test tubes were stab inoculated in same way as SIM media and incubated at 37°C for a week after which the findings were recorded. The test detects the ability of the bacterium to utilize citrate as the only carbon source which imparts blue in case of positive cases.

Lysine decarboxylase test: Test tubes with lysine broth were inoculated with pure overnight isolate followed by incubation at 37°C for 4 days. The findings were recorded after 96 hours of incubation. Absence of color change, i.e. the maintenance of purple color indicates the ability of the isolate to utilize the amino acid lysine and produce alkaline PH in the medium.

Sugar or carbohydrate fermentation tests: Conventional biochemical tests comprising four alcohols (dulcitol, mannitol, inositol and sorbitol) and two sugars (rhaminose and xylose) were used for adequate presumptive identification of E. tarda . Phenol red basal broth in Durham tubes containing 1% dulcitol, mannitol, sorbitol, inositol, xylose and rhamnose was prepared and inoculated with the isolates. The inoculates were then incubated at 37°C for 24 hours after which the results were recorded. Bacterial isolates with consistent characteristics of E. tarda based on primary and secondary identification criteria were considered presumptively as Edwardsiella tarda fish isolates and preserved in 15% glycerol (V/V) at -20°C for further characterization studies.

DATA ANALYSIS

Descriptive statistics such as proportions and frequency were employed in summarizing the data. Chi-square test of independence was employed in comparing the prevalence/occurrence of E. tarda infection with respect to site, sex, fish species and organ of isolation. A confidence interval of 95% was used to interpret the statistical association and significance was considered when P-value is less than 0.05.

Results

Of the total of 372 tissue samples comprising kidney, liver and intestine collected from 124 fish, E. tarda was isolated from 12 tissue samples (8 from liver, 3 from intestine and 1 from kidney). The isolates appeared as small punctuate grayish white colonies on xylose lysine deoxycholate agar after 24 hrs of incubation at 37°C. Except few of the isolates, most showed typical characteristics of E. tarda isolated elsewhere, which were gram negative short rods, motile, catalase positive and oxidase negative. In biochemical tests, these typical isolates were positive for indole, H2S production, and lysine decarboxylase and unable to utilize Simmon's citrate and the different sugars used in this study. However, some of the isolates showed variation from the typical characteristics. One isolate was negative for indole and able to utilize Simmon's citrate while the remaining was able to ferment mannitol, rhaminose, xylose and inositol. Although these isolates divert in some of the biochemical tests from the typical characteristics expected of E. tarda they were considered as variants of E. tarda strain due to the fact that the cultural characteristics, growth requirements and the different tests suggested the isolates to be within E. tarda species.

The major species of Edwardsiella those infect fish are E. tarda and E.ictaluri . E. tarda infects fish and other animals including human being, while E. Ictaluri being a pathogen of fish only. In this study, E. tarda were isolated from intestine, kidney and liver of fish indicating that the bacterium is a potential threat to aquaculture as well as to public health. The organism has been isolated from several sources previously such as from intestine of fish, humans faeces with sporadic cases of diarrhea and from dressed fish samples.

Concerning the morphological and biochemical characteristics of Edwardsiella tarda isolates, that showed typical characteristics, the results were consistent with those reported previously. The finding of one isolate negative for indole production and positive in Simmons citrate test, however, indicates atypical strain which was also reported in previous studies where variation in Simmons citrate utilization and indole production was reported among E. tardastrains.

The present study showed two isolates were found non motile and this fact is matched with that of Okuda et al.. Although most of the phenotypic characteristics of the isolates were similar as claimed by, some isolates showed, however, variation in some of the biochemical tests particularly in the utilization of sugars which included mannitol, rhamnose, xylose, inositol. The finding of such variation contradicts with the study of Baya et al. where no variation was observed with respect to these biochemical tests among fourty four E tarda isolates studied. Variation among E. tardaisolates, however, was reported with respected to utilization of rhaminose, mannitol.

The occurrence of variation in phenotypic characteristics among E. tarda isolates may due to the presence or absence of plasmid that control metabolic activities. Generally, the significance of the incidence of Edwardsiella tarda in catfish and tilapia couldn't be substantiated phenotypic characteristics of E. tarda.

Nowadays, more rapid, relatively accurate and simple molecular based identification techniques have been employed unlike the traditional phenotypic identification methods which rely on bacterial morphology and biochemical characteristics, Nucleic acid probes and the polymerase chain reaction (PCR) have been developed for the identification of pathogens of aquatic animals including E. tarda . Molecular methods such as PCR may sometimes yield false positive results as they can be highly subject to laboratory contamination.. Polymerase chain reaction based diagnosis of Edwardsiella tarda infection in blood samples of oyster toad fish and more sensitive real time PCR methods were reported as the major way in identifying the genes responsible for virulence of Edwardsiellosis .

Loop- mediated isothermal amplification (LAMP) is another rapid and sensitive method used for the diagnosis of Edwardsiellosis.

A fluorescence in situ hybridization (FISH) technique using twenty four mer oligonucleotide probe has also been used for detection of bacterial cells belonging to Enterobacteriaceae including E. tarda without giving false positive reaction.The use of higher performance capillary electrophoresis (HPCE) has also been reported as another technique which identifies, separate and quantifies intact bacteria. They identified E. tarda in fish and traced the bacteria in less than ten minutes after injection in to fish fluid using blue light emitting diode induced fluorescence and a cell permeable green nucleic acid strain.

The absence of significant differences in the occurrence of E. tarda between males and females indicates that both sexes are equally susceptible to the bacterium. This is an agreement several works where both sexes are equal chance of being infected with E. tarda. The significant differences in the rate of isolation of Edwardsiella tarda between the study lakes may be attributed to differences in the nutritional status of the fish, the environmental condition (Salinity and bacterial load of the water), water quality, changes in temperature, PH and fluctuation in dissolved oxygen which are believed to affect the occurrence of E. tarda infection. Although, Edwardsiella tarda affect intestine, liver and kidney of catfish and tilapia, the highest percentage of the pathogen was isolated from liver. This could be due to the metabolic activities of the organs.

In conclusion, Edwardsiella tarda is one of the most important bacterial diseases among Oreochromis niloticus in fish sampled from Lake Zeway and Langanoo. The severity of Edwardsiella tarda may have immune suppressive effect which proved by lymphoid depletion induced in spleen.

CONCLUSION AND RECOMMENDATIONS

Edwardsiellosis is the most important bacterial disease causing severe economic loss and hindrance in aqua farming. Apart from veterinary health importance, Edwardsiella tarda has also public health significance in people engaged in fishery industry and those depend on fish products for their annual income. The isolation of Edwarsiella tarda from wild fish population of Lakes Zeway and Langanoo destined for human consumption is, therefore, indicates that E. tarda is a potential threat of both the fishery sector/aquaculture and public health. The finding of certain isolates that divert in their biochemical characteristics warrants further investigation using more advanced methods of bacteria characterization. Generally, assessment of environmental condition, management and other stress factors enhancing the occurrence, distribution and severity of Edwardsiella tarda in fish is essential to design an effective disease control and prevention. Since the current state of knowledge of E. tarda infection in fish and humans in Ethiopia is almost nil, further study on the epidemiology of E. tarda in different hosts and environments as well as comprehensive information on the strains involved should be established for better fish productivity and public health.

10

AVIAN INFLUENZA

Avian influenza, known informally as avian flu or bird flu, is a variety of influenza caused by viruses adapted to birds. The type with the greatest risk is highly pathogenic avian influenza (HPAI). Bird flu is similar to swine flu, dog flu, horse flu and human flu as an illness caused by strains of influenza viruses that have adapted to a specific host. Out of the three types of influenza viruses (A, B, and C), influenza A virus is a zoonotic infection with a natural reservoir almost entirely in birds. Avian influenza, for most purposes, refers to the influenza A virus.

Avian influenza or bird flu refers to the disease caused by infection with avian (bird) influenza (flu) Type A viruses. These viruses naturally spread among wild aquatic birds worldwide and can infect domestic poultry and other bird and animal species. Bird flu viruses do not normally infect humans. However, sporadic human infections with bird flu viruses have occurred. The links below offer more information about bird flu.

Though influenza A is adapted to birds, it can also stably adapt and sustain person-to-person transmission. Recent influenza research into the genes of the Spanish flu virus shows it to have genes adapted from both human and avian strains. Pigs can also be infected with human, avian, and swine influenza viruses, allowing for mixtures of genes (reassortment) to create a new virus, which can cause an antigenic shift to a new influenza A virus subtype which most people have little to no immune protection against.

Avian influenza strains are divided into two types based on their pathogenicity: high pathogenicity (HP) or low pathogenicity (LP). The most well-known HPAI strain, H5N1, was first isolated from a farmed goose in Guangdong Province, China in 1996, and also has low pathogenic strains found in North America. Companion birds in captivity are unlikely to contract the virus and there has been no report of a companion bird with avian influenza since 2003. Pigeons can contract avian strains, but rarely become ill and are incapable of transmitting the virus efficiently to humans or other animals.

Between early 2013 and early 2017, 916 lab-confirmed human cases of H7N9 were reported to the World Health Organization (WHO). On 9 January 2017, the National Health and Family Planning Commission of China reported to the WHO 106 cases of H7N9 which occurred from late November through late December, including 35 deaths, 2 potential cases of human-to-human transmission, and 80 of these 106 persons stating that they have visited live poultry markets. The cases are reported from Jiangsu (52), Zhejiang (21), Anhui (14), Guangdong (14), Shanghai (2), Fujian (2) and Hunan (1). Similar sudden increases in the number of human cases of H7N9 have occurred in previous years during December and January.

The most widely quoted date for the beginning of recorded history of avian influenza (initially known as fowl plague) was in 1878 when it was differentiated from other diseases that caused high mortality rates in birds. Fowl plague, however, also included Newcastle disease until as recently as the 1950s. Between 1959 and 1995, there were 15 recorded occasions of the emergence of HPAI viruses in poultry, but losses were minimal. Between 1996 and 2008 however, HPAI outbreaks in poultry have occurred at least 11 times and 4 of these outbreaks have involved millions of birds.

In the 1990s, the world's poultry population grew 76% in developing countries and 23% in developed countries, contributing to the increased prevalence of avian influenza. Before the 1990s, HPAI caused high mortality in poultry, but infections were sporadic and contained. Outbreaks have become more common due to the high density and frequent movement of flocks from intensive poultry production.[

Influenza A/H5N1 was first isolated from a goose in China in 1996. Human infections were first reported in 1997 in Hong Kong. Since 2003, more than 700 human cases of Asian HPAI H5N1 have been reported to the WHO, primarily from 15 countries in Asia, Africa, the Pacific, Europe, and the Middle East, though over 60 countries have been affected.

GENETICS

Genetic factors in distinguishing between "human flu viruses" and "avian flu viruses" include:

PB2 (RNA polymerase)

Amino acid (or residue) position 627 in the PB2 protein encoded by the PB2 RNA gene. Until H5N1, all known avian influenza viruses had a Glu at position 627, while all human influenza viruses had a Lys.

HA (hemagglutinin)

Avian influenza HA viruses bind alpha 2-3 sialic acid receptors, while human influenza HA viruses bind alpha 2-6 sialic acid receptors. Swine influenza viruses have the ability to bind both types of sialic acid receptors. Hemagglutinin is the major antigen of the virus against which neutralizing antibodies are produced, and influenza virus epidemics are associated with changes in its antigenic structure. This was originally derived from pigs, and should technically be referred to as "pig flu".

The evolution of avian influenza virus has been influenced by genetic variation in the virus population due to genome segment reassortment and mutation. Also homologous recombination occurs in viral genes, suggesting that genetic variation generated by homologous recombination has also played a role in driving the evolution of the virus and potentially has affected virulence and host range.

Subtypes

There are many subtypes of avian influenza viruses, but only some strains of five subtypes have been known to infect humans: H5N1, H7N3, H7N7, H7N9, and H9N2. At least one person, an elderly woman in Jiangxi Province, China, died of pneumonia in December 2013 from the H10N8 strain. She was the first human fatality confirmed to be caused by that strain.

Most human cases of the avian flu are a result of either handling dead infected birds or from contact with infected fluids. It can also be spread through contaminated surfaces and droppings. While most wild birds have only a mild form of the H5N1 strain, once domesticated birds such as chickens or turkeys are infected, H5N1 can potentially become much more deadly because the birds are often in close contact. H5N1 is a large threat in Asia with infected poultry due to low hygiene conditions and close quarters. Although it is easy for humans to contract the infection from birds, human-to-human transmission is more difficult without prolonged contact. However, public health officials are concerned that strains of avian flu may mutate to become easily transmissible between humans.

Spreading of H5N1 from Asia to Europe is much more likely caused by both legal and illegal poultry trades than dispersing through wild bird migrations, being that in recent studies, there were no secondary rises in infection in Asia when wild birds migrate south again from their breeding grounds. Instead, the infection patterns followed transportation such as railroads, roads, and country borders, suggesting poultry trade as being much more likely. While there have been strains of avian flu to exist in the United States, they have been extinguished and have not been known to infect humans.

MODE OF TRANSMISSION

Avian influenza is most often spread by contact between infected and healthy birds, though can also be spread indirectly through contaminated equipment. The virus is found in secretions from the nostrils, mouth, and eyes of infected birds as well as their droppings. HPAI infection is spread to people often through direct contact with infected poultry, such as during slaughter or plucking.

Though the virus can spread through airborne secretions, the disease itself is not an airborne disease. Highly pathogenic strains spread quickly among flocks and can destroy a flock within 28 hours; the less pathogenic strains may affect egg production but are much less deadly.

Although it is possible for humans to contract the avian influenza virus from birds, human-to-human contact is much more difficult without prolonged contact. However, public health officials are concerned that strains of avian flu may mutate to become easily transmissible between humans. Some strains of avian influenza are present in the intestinal tract of large numbers of shore birds and water birds, but these strains rarely cause human infection.

Five manmade ecosystems have contributed to modern avian influenza virus ecology: integrated indoor commercial poultry, range-raised commercial poultry, live poultry markets, backyard and hobby flocks, and bird collection and trading systems including cockfighting. Indoor commercial poultry has had the largest impact on the spread of HPAI, with the increase in HPAI outbreaks largely the result of increased commercial production since the 1990s.

Village poultry

In the early days of the HPAI H5N1 pandemic, village poultry and their owners were frequently implicated in disease transmission. Village poultry, also known as backyard and hobby flocks, are small flocks raised under extensive conditions and often allowed free range between multiple households. However, research has shown that these flocks pose less of a threat than intensively raised commercial poultry with homogenous genetic stock and poor biosecurity. Backyard and village poultry also do not travel great distances compared to transport of intensively raised poultry and contribute less to the spread of HPAI. This initial implication of Asian poultry farmers as one broad category presented challenges to prevention recommendations as commercial strategies did not necessarily apply to backyard poultry flocks.

H5N1

The highly pathogenic influenza A virus subtype H5N1 is an emerging avian influenza virus that is causing global concern as a potential pandemic threat. It is often referred to simply as "bird flu" or "avian influenza", even though it is only one of many subtypes.

H5N1 has killed millions of poultry in a growing number of countries throughout Asia, Europe, and Africa. Health experts are concerned that the coexistence of human flu viruses and avian flu viruses (especially H5N1) will provide an opportunity for genetic material to be exchanged between species-specific viruses, possibly creating a new virulent influenza strain that is easily transmissible and lethal to humans. The mortality rate for humans with H5N1 is 60%.

Since the first human H5N1 outbreak occurred in 1997, there has been an increasing number of HPAI H5N1 bird-to-human transmissions, leading to clinically severe and fatal human infections. Because a significant species barrier exists between birds and humans, the virus does not easily spread to humans, however some cases of infection are being researched to discern whether human-to-human transmission is occurring. More research is necessary to understand the pathogenesis and epidemiology of the H5N1 virus in humans. Exposure routes and other disease transmission characteristics, such as genetic and immunological factors that may increase the likelihood of infection, are not clearly understood.

The first known transmission of H5N1 to a human occurred in Hong Kong in 1997, when there was an outbreak of 18 human cases; 6 deaths were confirmed. None of the infected people worked with poultry. After culling all of the poultry in the area, no more cases were diagnosed. In 2006, the first human-to-human transmission likely occurred when 7 members of a family in Sumatra became infected after contact with a family member who had worked with infected poultry.

Although millions of birds have become infected with the virus since its discovery, 359 people have died from H5N1 in twelve countries according to World Health Organization reports as of August 10, 2012.

The H5N1 outbreak in Thailand caused massive economic losses, especially among poultry workers. Infected birds were culled and slaughtered. The public lost confidence with the poultry products, thus decreasing the consumption of chicken products. This also elicited a ban from importing countries. There were, however, factors which aggravated the spread of the virus, including bird migration, cool temperature (increases virus survival) and several festivals at that time.

A mutation in the virus was discovered in two Guangdong patients in February 2017 which rendered it more deadly to chickens, inasmuch as it could infect every organ; the risk to humans was not increased, however.

CONTROVERSIAL RESEARCH

A study published in 2012 in Science Magazine reported on research findings that allowed for the airborne transmission of H5N1 in laboratory ferrets. The study identified the 5 mutations necessary for the virus to become airborne and immediately sparked controversy over the ethical implications of making such potentially dangerous information available to the general public. The study was allowed to remain available in its entirety, though it remains a controversial topic within the scientific community.

The study in question, however, created airborne H5N1 via amino acid substitutions that largely mitigated the devastating effects of the disease. This fact was underscored by the 0% fatality rate among the ferrets infected via airborne transmission, as well as the fundamental biology underlying the substitutions. Flu viruses attach to host cells via the hemagluttinin proteins on their envelope. These hemagluttinin proteins bind to sialic acid receptors on host cells, which can fall into two categories. The sialic acid receptors can be either 2,3 or 2,6-linked, with the species of origin largely deciding receptor preference. In influenzas of avian origin 2,3-linkage is preferred, vs. influenzas of human origin in which 2,6-linkage is preferable. 2,3-linked SA receptors in humans are found predominantly in the lower respiratory tract, a fact that is the primary foundation for the deadliness of avian influenzas in humans, and also the key to their lack of airborne transmission. In the study that created an airborne avian influenza among ferrets it was necessary to switch the receptor preference of the host cells to those of 2,6-linkage, found predominantly in humans' upper respiratory tract, in order to create an infection that could shed aerosolized virus particles. Such an infection, however, must occur in the upper respiratory tract of humans, thus fundamentally undercutting the fatal trajectory of the disease.

H7N9

Influenza A virus subtype H7N9 is a novel avian influenza virus first reported to have infected humans in 2013 in China. Most of the reported cases of human infection have resulted in severe respiratory illness. In the month following the report of the first case, more than 100 people had been infected, an unusually high rate for a new infection; a fifth of those patients had died, a fifth had recovered, and the rest remained critically ill. The WHO has identified H7N9 as "...an unusually dangerous virus for humans." As of June 30, 133 cases have been reported, resulting in the deaths of 43.

Research regarding background and transmission is ongoing. It has been established that many of the human cases of H7N9 appear to have a link to live bird markets. As of July 2013, there had been no evidence of sustained human-to-human transmission, however a study group headed by one of the world's leading experts on avian flu reported that several instances of human-to-human infection were suspected. It has been reported that H7N9 virus does not kill poultry, which will make surveillance much more difficult. Researchers have commented on the unusual prevalence of older males among H7N9-infected patients. While several environmental, behavioral, and biological explanations for this pattern have been proposed, as yet, the reason is unknown. Currently no vaccine exists, but the use of influenza antiviral drugs known as neuraminidase inhibitors in cases of early infection may be effective.

The number of cases detected after April fell abruptly. The decrease in the number of new human H7N9 cases may have resulted from containment measures taken by Chinese authorities, including closing live bird markets, or from a change in seasons, or possibly a combination of both factors. Studies indicate that avian influenza viruses have a seasonal pattern, thus it is thought that infections may pick up again when the weather turns cooler in China.

In the four years from early 2013 to early 2017, 916 lab-confirmed human cases of H7N9 were reported to the WHO.

On 9 January 2017, the National Health and Family Planning Commission of China reported to the WHO 106 cases which occurred from late November through December. 29, 2016. The cases are reported from Jiangsu (52), Zhejiang (21), Anhui (14), Guangdong (14), Shanghai (2), Fujian (2) and Hunan (1). 80 of these 106 persons have visited live poultry markets. Of these cases, there have been 35 deaths. In two of the 106 cases, human-to-human transmission could not be ruled out.

Affected prefectures in Jiangsu province closed live poultry markets in late December 2016, whereas Zhejiang, Guangdong and Anhui provinces went the route of strengthening live poultry market regulations. Travellers to affected regions are recommended to avoid poultry farms, live bird markets, and surfaces which appear to be contaminated with poultry feces. Similar sudden increases in the number of human cases of H7N9 have occurred in previous years during December and January.

DOMESTIC ANIMALS

Several domestic species have been infected with and shown symptoms of H5N1 viral infection, including cats, dogs, ferrets, pigs, and birds.

Birds

Attempts are made in the United States to minimize the presence of HPAI in poultry through routine surveillance of poultry flocks in commercial poultry operations. Detection of a HPAI virus may result in immediate culling of the flock. Less pathogenic viruses are controlled by vaccination, which is done primarily in turkey flocks (ATCvet codes: QI01AA23 (WHO) for the inactivated fowl vaccine, QI01CL01 (WHO) for the inactivated turkey combination vaccine).

Cats

Avian influenza in cats can show a variety of symptoms and usually lead to death. Cats are able to get infected by either consuming an infected bird or by contracting the virus from another infected cat.

Global impact

In 2005, the formation of the International Partnership on Avian and Pandemic Influenza was announced in order to elevate the importance of avian flu, coordinate efforts, and improve disease reporting and surveillance in order to better respond to future pandemics. New networks of laboratories have emerged to detect and respond to avian flu, such as the Crisis Management Center for Animal Health, the Global Avian Influenza Network for Surveillance, OFFLU, and the Global Early Warning System for major animal diseases. After the 2003 outbreak, WHO member states have also recognized the need for more transparent and equitable sharing of vaccines and other benefits from these networks. Cooperative measures created in response to HPAI have served as a basis for programs related to other emerging and re-emerging infectious diseases.

HPAI control has also been used for political ends. In Indonesia, negotiations with global response networks were used to recentralize power and funding to the Ministry of Health. In Vietnam policymakers, with the support of the Food and Agriculture Organization of the United Nations (FAO), used HPAI control to accelerate the industrialization of livestock production for export by proposing to increase the portion of large-scale commercial farms and reducing the number of poultry keepers from 8 to 2 million by 2010.

Bird Flu in 2020

By the end of 2020, several outbreaks of Avian flu of various varieties were reported in Europe. Since mid-October several European countries, including Belgium, Denmark, France, Germany, Ireland, the Netherlands, Sweden, and the United Kingdom have reported outbreaks of highly pathogenic avian influenza (HPAI) viruses, mostly in wild birds. Positive tests were also among poultry and captive birds. According to a report by the European Centre for Disease Prevention and Control (ECDC), three varieties of HPAI viruses were found, A(H5N8), A(H5N5) and A(H5N1), with H5N8 being the most commonly found. In Germany, 29,000 chickens were killed to halt the spread of H5N8. In Belgium, H5N5 was found on a poultry farm according to the World Organization for Animal Health (OIE). The outbreak was reported in Menen, near the border with France, and killed 600 birds and the culling of an additional 151,000 chickens from the flock.

United States 2014–15 outbreak

From 2014 through 2015, United States poultry and egg producers experienced the largest outbreak of H5N2 in recorded history with approximately 51 million birds depopulated to control the spread of the disease. From May to June 2015, 25 million birds were culled, equating to 409,836 birds per day, or 284 birds per minute. In total, the 2014-2015 H5N2/H5N8 outbreak cost US$879 million in public expenditures and the United States egg and poultry industry more than US$3 billion to eradicate the disease from poultry production. This was the most costly United States HPAI outbreak to date.

Stigma

Backyard poultry production was viewed as "traditional Asian" agricultural practices that contrasted with modern commercial poultry production and seen as a threat to biosecurity. Backyard production appeared to hold greater risk than commercial production due to lack of biosecurity and close contact with humans, though HPAI spread in intensively raised flocks was greater due to high density rearing and genetic homogeneity. Asian culture itself was blamed as the reason why certain interventions, such as those that only looked at placed-based interventions, would fail without looking for a multifaceted solutions.

Indonesia

Press accounts of avian flu in Indonesia were seen by poultry farmers as conflating suspected cases while the public did see the accounts as informative, though many became de-sensitized to the idea of impending danger or only temporarily changed their poultry-related behavior. Rumors also circulated in Java in 2006. These tended to focus on bird flu being linked to big businesses in order to drive small farmers out of the market by exaggerating the danger of avian influenza, avian flu being introduced by foreigners to force Indonesians to purchase imported chicken and keep Indonesian chicken off the world market, and the government using avian flu as a ploy to attract funds from wealthy countries. Such rumors reflected concerns about big businesses, globalization, and a distrust of the national government in a country where "the amount of decentralization here is breathtaking" according to Steven Bjorge, a WHO epidemiologist in Jakarta in 2006.

In the context a decentralized national government that the public did not completely trust, Indonesian Health Minister Siti Fadilah Supari announced in December 2006 that her government would no longer be sharing samples of H5N1 collected from Indonesian patients. This decision came as a shock to the international community as it disrupted the Global Influenza Surveillance Network (GISN) coordinated by the WHO for managing seasonal and pandemic influenza. GISN is based on countries sharing virus specimens freely with the WHO which assesses and eventually sends these samples to pharmaceutical companies in order to produce vaccines that are sold back to these countries. Though this was initially seen as an attempt to protect national sovereignty at all costs, it was instead used for a domestic political struggle. Prior to Indonesia's dispute with the GISN, the Ministry of Health, already weak due to the decentralized nature the government, was experiencing further leakage of funding to state and non-state agencies due to global health interventions. By reasserting control over public health issues and funding by setting itself up as the sole Indonesian representative to the WHO, the Ministry of Health made itself a key player in the management of future international funds relating vaccine production and renegotiated benefits from global surveillance networks.

ECONOMIC

Approximately 20% of the protein consumed in developing countries come from poultry. In the wake of the H5N1 pandemic, millions of poultry were killed. In Vietnam alone, over 50 million domestic birds were killed due to HPAI infection and control attempts. A 2005 report by the FAO totaled economic losses in South East Asia around US$10 billion. This had the greatest impact on small scale commercial and backyard producers relative to total assets compared to industrial chains which primarily experience temporary decreases in exports and loss of consumer confidence. Some governments did provide compensation for culled poultry, it was often far below market value (close to 30% of market value in Vietnam), while others such as Cambodia provide no compensation to farmers at all.

As poultry serves as a source of food security and liquid assets, the most vulnerable populations were poor small scale farmers. The loss of birds due to HPAI and culling in Vietnam led to an average loss of 2.3 months of production and US$69–108 for households where many have an income of $2 a day or less. The loss of food security for vulnerable households can be seen in the stunting of children under 5 in Egypt. Women are another population at risk as in most regions of the world, small flocks are tended to by women. Widespread culling also resulted in the decreased enrollment of girls in school in Turkey.

PREVENTION

People who do not regularly come into contact with birds are not at high risk for contracting avian influenza. Those at high risk include poultry farm workers, animal control workers, wildlife biologists, and ornithologists who handle live birds. Organizations with high-risk workers should have an avian influenza response plan in place before any cases have been discovered. Biosecurity of poultry flocks is also important for prevention. Flocks should be isolated from outside birds, especially wild birds, and their waste; vehicles used around the flock should be regularly disinfected and not shared between farms; and birds from slaughter channels should not be returned to the farm.

With proper infection control and use of personal protective equipment (PPE), the chance for infection is low. Protecting the eyes, nose, mouth, and hands is important for prevention because these are the most common ways for the virus to enter the body. Appropriate personal protective equipment includes aprons or coveralls, gloves, boots or boot covers, and a head cover or hair cover. Disposable PPE is recommended. An N-95 respirator and unvented/indirectly vented safety goggles are also part of appropriate PPE. A powered air purifying respirator (PAPR) with hood or helmet and face shield is also an option.

Proper reporting of an isolated case can help to prevent spread. The Centers for Disease Control and Prevention (US) recommendation is that if a worker develops symptoms within 10 days of working with infected poultry or potentially contaminated materials, they should seek care and notify their employer, who should notify public health officials.

For future avian influenza threats, the WHO suggests a 3 phase, 5 part plan.

- Phase 1: Pre-pandemic
- Reduce opportunities for human infection
- Strengthen the early warning system
- Phase 2: Emergence of a pandemic virus
- Contain or delay spread at the source
- Phase 3: Pandemic declared and spreading internationally
- Reduce morbidity, mortality, and social disruption
- Conduct research to guide response measures

Vaccines for poultry have been formulated against several of the avian H5N1 influenza varieties. Control measures for HPAI encourage mass vaccinations of poultry though The World Health Organization has compiled a list of known clinical trials of pandemic influenza prototype vaccines, including those against H5N1. In some countries still at high risk for HPAI spread, there is compulsory strategic vaccination though vaccine supply shortages remain a problem.

For village poultry farmers

During the initial response to H5N1, a one size fits all recommendation was used for all poultry production systems, though measures for intensively raised birds were not necessarily appropriate for extensively raised birds. When looking at village-raised poultry, it was first assumed that the household was the unit and that flocks did not make contact with other flocks, though more effective measures came into use when the epidemiological unit was the village.

Recommendations involve restructuring commercial markets to improve biosecurity against avian influenza. Poultry production zoning is used to limit poultry farming to specific areas outside of urban environments while live poultry markets improve biosecurity by limiting the number of traders holding licenses and subjecting producers and traders to more stringent inspections. These recommendations in combination with requirements to fence and house all poultry, and to limit free ranging flocks, will eventually lead to fewer small commercial producers and backyard producers, costing livelihoods as they are unable to meet the conditions needed to participate.

A summary of reports to the World Organisation for Animal Health in 2005 and 2010 suggest that surveillance and under-reporting in developed and developing countries is still a challenge. Often, donor support can focus on HPAI control alone, while similar diseases such as Newcastle disease, acute fowl cholera, infectious laryngotracheitis, and infectious bursal disease still affect poultry populations. When HPAI tests come back negative, a lack of funded testing for differential diagnoses can leave farmers wondering what killed their birds.

Since traditional production systems require little investment and serve as a safety net for lower income households, prevention and treatment can be seen as less cost-effective than letting poultry die. Effective control not only requires prior agreements to be made with relevant government agencies, such as seen with Indonesia, they must also not unduly threaten food security.

Culling

Culling is used in order to decrease the threat of avian influenza transmission by killing potentially infected birds. The FAO manual on HPAI control recommends a zoning strategy which begins with the identification of an infected area (IA) where sick or dead birds have tested positive. All poultry in this zone are culled while the area 1 to 5 km from the outer boundary of the IA is considered the restricted area (RA) placed under strict surveillance. 2 to 10 km from the RA is the control area (CA) that serves as a buffer zone in case of spread. Culling is not recommended beyond the IA unless there is evidence of spread.

The manual, however, also provides examples of how control was carried out between 2004 and 2005 to contain H5N1 where all poultry was to be stamped out in a 3 km radius beyond the infected point and beyond that a 5 km radius where all fowl was to be vaccinated. This culling method was indiscriminate as a large proportion of the poultry inside these areas were small backyard flocks which did not travel great enough distances to carry infection to adjacent villages without human effort and may have not been infected at all. Between 2004 and 2005, over 100 million chickens were culled in Asia to contain H5N1.

The risk of mass culling of birds and the resulting economic impact led to farmers who were reluctant to report sick poultry. The culls often preempted actual lab testing for H5N1 as avian flu policy justified sacrificing poultry as a safeguard against HPAI spread. In response to these policies, farmers in Vietnam between 2003 and 2004 became more and more unwilling to surrender apparently healthy birds to authorities and stole poultry destined for culls as it stripped poultry of their biosocial and economic worth. By the end of 2005, the government implemented a new policy that targeted high-risk flock in the immediate vicinity of infected farms and instituted voluntary culling with compensation in the case of a local outbreak.

Not only did culling result in severe economic impacts especially for small scale farmers, culling itself may be an ineffective preventative measure. In the short-term, mass culling achieves its goals of limiting the immediate spread of HPAI, it has been found to impede the evolution of host resistance which is important for the long-term success of HPAI control. Mass culling also selects for elevated influenza virulence and results in the greater mortality of birds overall. Effective culling strategies must be selective as well as considerate of economic impacts to optimize epidemiological control and minimize economic and agricultural destruction.

People-poultry relations

Prevention and control programs must take into account local understandings of people-poultry relations. In the past, programs that have focused on singular, place-based understandings of disease transmission have been ineffective. In the case of Northern Vietnam, health workers saw poultry as commodities with an environment that was under the control of people. Poultry existed in the context of farms, markets, slaughterhouses, and roads while humans were indirectly the primary transmitters of avian flu, placing the burden of disease control on people. However, farmers saw their free ranging poultry in an environment dominated by nonhuman forces that they could not exert control over. There were a host of nonhuman actors such as wild birds and weather patterns whose relationships with the poultry fostered the disease and absolved farmers of complete responsibility for disease control.

11

AEROMONOUS HYDROPHILA BACTERIUM

Aeromonas hydrophila is a heterotrophic, Gram-negative, rod-shaped bacterium mainly found in areas with a warm climate. This bacterium can be found in fresh or brackish water. It can survive in aerobic and anaerobic environments, and can digest materials such as gelatin and hemoglobin. A. hydrophila was isolated from humans and animals in the 1950s. It is the best known of the species of Aeromonas. It is resistant to most common antibiotics and cold temperatures and is oxidase- and indole-positive. Aeromonas hydrophila also has a symbiotic relationship as gut flora inside of certain leeches, such as Hirudo medicinalis.

STRUCTURE

Aeromonas hydrophila bacteria are Gram-negative, straight rods with rounded ends (bacilli to coccibacilli shape) usually from 0.3 to 1.0 μm in width and 1.0 to 3.0 μm in length. They can grow at temperatures as low as 4 °C. These bacteria are motile by a polar flagellum.

Pathology

Because of its structure, it is very toxic to many organisms. When it enters the body of its victim, it travels through the bloodstream to the first available organ. It produces aerolysin, a cytotoxic enterotoxin that can cause tissue damage. A. hydrophila, A. caviae, and A. sobria are all considered to be opportunistic pathogens, meaning they rarely infect healthy individuals. A. hydrophila is widely considered a major fish and amphibian pathogen, and its pathogenicity in humans has been recognized for decades. The genomic insights of aeromonas could be a stepping stone into understanding of them

Pathogenic mechanism

The pathogenicity of Aeromonas species was believed to be mediated by a number of extracellular proteins such as aerolysin, lipase, chitinase, amylase, gelatinase, hemolysins, and enterotoxins. However, the pathogenic mechanisms are unknown. The recently proposed type-III secretion system (T3SS) has been linked to Aeromonas pathogenesis.

T3SS is a specialized protein secretion machinery that exports virulence factors directly to host cells. These factors subvert normal host cell functions to the benefit of invading bacteria. In contrast to the general secretory pathway, the T3SS is triggered when a pathogen comes in contact with host cells. ADP-ribosylation toxin is one of the effector molecules secreted by several pathogenic bacteria and translocated through the T3SS and delivered into the host cytoplasm, which leads to interruption of the NF-?B pathway, cytoskeletal damage, and apoptosis. This toxin has been characterized in A. hydrophila (human diarrhoeal isolate), A. salmonicida (fish pathogen), and A. jandaei GV17, a pathogenic strain that can cause disease both in humans and fish.

Occurrence of exposure

Aeromonas hydrophila infections occur most often during sexual changes, stressors, changes in temperature, in contaminated environments, and when an organism is already infected with a virus or another bacterium. It can also be ingested through food products contaminated with the bacterium, such as seafood, meats, and even certain vegetables such as sprouts. It can also be transmitted by leeches.

DIAGNOSIS OF AEROMONAS HYDROPHILA

Infection In Fish

Motile aeromonads of the Aeromonas hydrophila complex cause a haemorrhagic septicaemia in fish (Bullock et al., 1971; Egusa, 1978; Schäperclaus et al., 1992). This bacterium has been observed in numerous species of freshwater fish and occasionally in marine fish and in amphibians, reptiles, cattle and humans throughout the world (Bullock et al., 1971; Khardori and Fainstein, 1988). However, the most significant diseases occur in cultured freshwater fish. The bacterium is distributed widely in fresh water and bottom sediments containing organic material, as well as in the intestinal tract of fish (Aoki, 1974; Egusa, 1978; Hazen et al., 1978; Seidler et al., 1980; Kaper et al., 1981; Van der Kooij and Hijnen, 1988; Sugita et al., 1994, Dumontet et al., 1996).

Infectious abdominal dropsy in common carp has been attributed to the A. hydrophila group (Aeromonas punctata) and was first described by Schäperclaus (1930), who reported on this condition in cultured, wild and stocked carp in central and eastern Europe. The causative agent has since been shown to be rhabdovirus carpio (spring viraemia of carp) (Fijan, 1972; Wolf, 1988). During the 1960s, outbreaks of red fin disease, caused by A. hydrophila, occurred frequently in cultured eels in Japan (Hoshina, 1962; Egusa, 1978), with concurrent infections by Saprolegnia parasitica (Egusa, 1978). Currently, only sporadic outbreaks of A. hydrophila occur in cultured eels. Aeromonas hydrophila is typically recognized as an opportunistic pathogen or secondary invader (Austin and Austin, 1987).

Conversely, there have been reports of A. hydrophila acting as a primary pathogen in fish. Isolates differ greatly in their pathogenicity with some strains being highly virulent and others non-virulent. Eddy (1960) and Kou (1972a) reported that non-virulent or weakly pathogenic strains did not produce gas and acetone from glucose. Wakabayashi et al. (1981) recognized common and identical biochemical characteristics in virulent strains, in particular the production of elastase and enzymes involved in lysis of Staphylococcus. These characteristics were identical to the A. hydrophila biovar. hydrophila, following classification by Popoff and Véron (1976). Economically, the disease attributed to these bacteria is of greatest importance in cultured freshwater fish. Recent advances in biochemistry, molecular biology and virulence factors associated with A. hydrophila have led to new understanding of this bacterial group.

Aeromonas hydrophila is associated with diseases mainly found in freshwater fish and amphibians, because these organisms live in aquatic environments. It is linked to a disease found in frogs called red leg, which causes internal, sometimes fatal hemorrhaging. When infected with A. hydrophila, fish develop ulcers, tail rot, fin rot, and hemorrhagic septicemia. Hemorrhagic septicaemia causes lesions that lead to scale shedding, hemorrhages in the gills and anal area, ulcers, exophthalmia, and abdominal swelling.

Aeromonashydrophila causes disease in fish known as "Hem-morhagicSepticemia", "Motile Aeromonas Septicemia", "Ulcer Disease", or "Red-Sore Disease". The many synonyms of this disease relate to the lesions caused by this bacterium which includes lesions of sep-ticemia when the bacteria or bacterial toxins are present within numerous organs within the fish, and ulcers of. the fish's skin. Aeromonashydro-phila is a ubiquitous gram-negative, motile, rod-shaped' bacterium which can be commonly isolated from fresh water ponds and is also a normal inhabitant of the gastrointestinal tract of fish. The disease primarily affects freshwater fish such as catfish and bass, and many species of tropical or ornamental fish.

Fish infected with Aeromonashydrophila may have many different clinical signs. These range from sudden death in otherwise healthy fish to inap-petence, swimming abnormalities, pale gills, bloat and skin ulcerations. The skin ulcers may occur at any site on the fish and often they are surrounded by a bright red rim of tissue. Because of the variability of these clinical signs, the diagnosis of this disease based only upon the clinical presentation of the fish is highly unreliable and may be economically disastrous to the fish producer.

With intensive fish farming systems, whether these systems are outdoor ponds or indoor aquaria and tanks, predisposing factors are primarily responsible for the precipitation of this disease. Stress is the single most important predisposing factor associated with this disease. Stress is due to poor management and/or poor water quality. Management factors include:

Nutrition, handling, transportation and over-crowding of fish. Water quality must be excellent to prevent this disease. This means that the dissolved oxygen (DO), pH, temperature and alkalinity of the water must be satisfactory, and that the ammonia, nitrite and COolevels must be kept to a minimurn.

Diagnosis

The gross signs of disease are haemorrhagic septicaemia and fin rot. The aetiological agent can be grown on brain-heart infusion medium, tryptosoy agar (tryptone soya agar), nutrient agar and MacConkey agar with incubation at 20-25°C for 24-48 h. Numerous selective media have been developed for the isolation and presumptive identification of Aeromonas hydrophila or motile aeromonads (Moyer, 1996), including Rimler-Shotts medium (Shotts and Rimler, 1973), modified peptone beef-extract glycogen agar (McCoy and Seidler, 1973), Rippey-Cabelli (membrane filter method (mA)) agar (Rippey and Cabelli, 1979), MacConkey's agar supplemented with trehalose (Kaper et al., 1981) and starch-ampicillin agar (Palumbo et al., 1985). Davis and Sizemore (1981) reported that Rimler and Shotts medium and Rippey-Cabelli agar were not suitable for A. hydrophila. Arcos et al. (1988) compared six media for selective isolation of A. hydrophila and showed that mA agar gave the best recovery rate and also an acceptable specificity, but its selectivity was low. An API-20E test strip is used widely for identification of the Enterobacteriaceae (Kaper et al., 1979). Toranzo et al. (1986) indicated that the importance of biochemical characteristics must be backed up by standardized testing.

For diagnosis, the aetiological agent is isolated on nutrient agar at 30°C from kidney of a fish with haemorrhagic septicaemia. Isolates are motile Gram-negative bacilli. Wakabongo et al. (1992) found that only four tests - aesculin hydrolysis, acetoin production, lysine decarboxylation and gas from glucose - were sufficient to distinguish A. hydrophila, A. caviae and A. sobria. A number of A. hydrophila-lytic bacteriopahges were isolated from river water, river mud, sewage and human clinical origin (e.g. stool, urine) and phage typing was attempted (Chow and Rouf, 1983; Demarta and Peduzzi, 1984; Altwegg et al., 1988; Merino et al., 1990; Fukuyama et al., 1991, 1992). A comprehensive phage-typing system for A. hydrophila will provide a useful tool for epizootiological and ecological studies. Aeromonas hydrophila has been identified by the gel-diffusion technique (Bullock, 1966), direct fluorescent antibody technique (Lewis and Allison, 1971), indirect fluorescent antibody technique (Lewis and Savage, 1972), immunoblotted sodium dodecylsulphate (SDS)-polyacrylamide gel electrophoresis (PAGE) (Mulla and Millership, 1993) and enzyme-linked immunosorbent assay (Merino et al., 1993). These methods are of limited value, because many different serological types of A. hydrophila are distributed in fish farms (Eddy, 1960; Bullock, 1966).

Lucchini and Altwegg (1992) differentiated ribotyping of restriction genomic DNAs of aeromonads, using different fragments of a 16S rDNA gene of E. coli as a probe and achieved identification of most Aeromonas strains to species level. This method is easier than DNA-DNA hybridization, because only minimal amounts of genomic DNA are needed and several strains can be analysed on a single gel. Pulsed-field gel electrophoresis is a rapid and discriminatory technique for the typing of A. hydrophila where a common origin of infection is suspected (Talon et al., 1996).

Deoxyribonucleic acid probe hybridization technology is now becoming available for the direct detection and identification of microorganisms. The colony hybridization technique (Grunstein and Hogness, 1975) can be applied easily to identify and count the causative organism. Fish-pathogenic bacteria, such as Vibrio anguillarum (Hirono et al., 1996) and Pasteurella piscicida (Zhao and Aoki, 1992a), can be detected by probe hybridization with their specific DNA nucleotide sequence or cryptic plasmid. However, there are many common DNA fragments between A. hydrophila and Aeromonas salmonicida (Miyata et al., 1995), which makes it unlikely that this probe technique will be successful for this species.

The sensitivity obtained using hybridization with non-radiolabelled probes is lower than with radiolabelled probes (Zhao and Aoki, 1992a). However, radiolabelled probes are generally unacceptable for use in diagnosis and further development of detection procedures based on non-radiolabelling is still required.

The polymerase chain reaction (PCR) (Mullis and Faloona, 1987) is useful for detecting fish pathogens from diseased fish and their environment. This technique is available for A. salmonicida (Miyata et al., 1996), Edwardsiella tarda (Aoki and Hirono, 1995), P. piscicida (Aoki et al., 1997) and V. anguillarum (Hirono et al., 1996). However, recently, Cascón et al. (1996) found a specific PCR primer set for the detection of A. hydrophila hybridization group 1.

The DNA fingerprinting method, AFLP (amplified fragment length polymorphism), is a valuable high-resolution genotype tool for classification of Aeromonas species (Huys et al., 1996).

Aeromonas hydrophila isolated from wild freshwater fish in Croatia.

Aeromonas hydrophila was recovered from fish living in lake Vrana on the Croatian island of Cres. The occurrence of the bacterium in the fish was assessed and related to gross signs of disease and findings at necropsy as a potential health hazard for fish. Isolated bacteria were subjected to morphological, physiological, biochemical and antibiotic susceptibility tests. A total of 26 A. hydrophila isolates were obtained. There was a clear seasonality, as no isolates were recovered in the summer months. Most of the isolates were sensitive to all the antimicrobials used in the study except novobiocin and penicillin G. Affected fish manifested haemorrhages over the skin, in the liver, kidney and swim bladder, spleen infarcts, fatty liver, ascitic fluid and swollen haemopoietic tissues. A. hydrophila does not appear to pose a major threat for the fish in the lake at present but under unfavourable and stressful conditions it could seriously compromise fish health.

Pathological findings of experimental Aeromonas hydrophila infection in Golden Mahseer

Introduction

Mahseers, belong to the family Cyprinidae, are well renowned for the excellent game as well as food fishes. Various aspects simultaneously considered towards the domestication of this fish species also include the study of fish health issues, preventive and remedial programmes for sustainable aquaculture. Experimental studies can be conducted in the control condition to study the pathological symptoms during bacterial infection. Considering the bacterial pathogens as a major constraint for aquaculture, the present study was carried out to standardize the bacterial concentration of A. hydrophilafor the experimental challenge of Golden Mahseer for the first time, and various clinical changes were also monitored during the infection.

Materials and Methods

A total of 140 live fishes were acclimatized to the laboratory conditions for experimental study. A constant water quality was monitored during the acclimation period and throughout the experiment. The test group was injected with A. hydrophila and various physiological, biochemical and tissue level changes were observed during the experiment in comparison to phosphate buffer saline (PBS) injected control group.

Results

LD50 value of A. hydrophila for Golden Mahseer was successfully standardized. Clinical signs including weakness, slower movement, swimming closer to the surface, fin haemorrhages and red patches at the gut region were observed. Enlargement of spleen followed by tissue necrosis along with signs of haemorrhagic septicaemia was also seen in infected fishes. The necrosis of hepatocytes was observed in the histological section of liver tissue. The bacterial infection increases the superoxide dismutase activity and cortisol level in Golden Mahseer.

Conclusion

The LD50 value of A. hydrophila for experimental challenge studies on Golden Mahseer is 1.74 × 105 cfu per 100 g of body weight. The symptoms of severe bacterial haemorrhage were observed. Necrosis of hepatocytes was observed in infected fishes. The superoxide dismutase activity and cortisol level also increased in infected fishes. A. hydrophila was confirmed to be the etiological agent which was re-isolated using spread plate method and confirmed by 16s rRNA sequencing.

Mahseers, belong to the family Cyprinidae, are well known for the excellent game as well as food fishes all over the world especially the South-Asian countries. In India, eight mahseer species are available among which Golden Mahseer, Tor putitora is considered the candidate species for hill aquaculture in the mid-Himalayan range. The artificial breeding, seed production and scientific management of Golden Mahseer are in high priorities. Scientific management including water quality, feed and health management are also equally important for sustainable aquaculture. However, the data regarding the health management, mainly diseases is least available for this important fish species.

Aquaculture is primarily affected by microbial pathogens especially of bacterial origin, wherein, Aeromonas hydrophilaand other Aeromonas are responsible for the majority of diseases. A. hydrophila is reported being primary bacterial pathogen affecting freshwater fishes like Clarias gariepinus , Labeo rohita , Sparus aurata , Magalobrama amblycephala . The severity of pathogenesis or disease outbreak can be moderate to high in fishes, but every type of disease results in deterioration of product and economic losses.

Aeromonad septicaemia is characterized by diverse pathological symptoms such as dermal ulceration, fin haemorrhages, fin rots, red sores, haemorrhages and necrosis of the visceral organs, etc. The acute form of the disease may result in fatal sepsis without any symptoms while chronic infections may show the symptoms like hemorrhagic septicaemia with ulceration, inflammation, and dermal lesions. Liver and kidney tissues are the primary targets of bacterial accumulation and sepsis The liver may become pale and show greyish to greenish coloration with foci while kidney may engorge and become friable. Despite the knowledge of severity and symptoms of the Aeromonad septicaemia in other species, factors such as host-pathogen interaction, temperature requirement of the bacteria, course of pathogenesis, and immunity to pathogenesis may vary for fish species.

The present study was aimed to examine the effect of Aeromonas hydrophila infection on Golden Mahseer. Various macroscopic and microscopic changes in tissues and stress level was investigated to understand the fish response to bacterial infection. As the liver tissue is the primary organ affected by bacterial infection, microscopic changes in hepatic cells were also examined to ascertain the level of cell necrosis due to pathogenesis.

Materials and Methods

Ethical approval

The institutional ethical committee approved the protocols for maintenance, handling during experiments, and sacrificing of the fishes.

Experimental fish

Live fishes (size=33.32 ± 5.96 g; 146 ± 15.17 mm) were collected from Kosi River, Ramnagar (29.40°N 79.12°E) using a cast net. The fishes were initially acclimatized in ponds for three weeks and then in fibre reinforced plastic tanks (experimental tanks) for two weeks of 2000 L capacity before the experiment. Fishes were fed twice, daily with a formulated diet based on 5% of the body weight. Constant aeration and water flow were maintained in experimental tanks. Water temperature and physicochemical parameters were recorded at regular intervals using multi-parameter auto-analyser (Hach , Colorado, US) and colorimetric methods.

Bacterial strain and determination of LD50

The bacterial strain (originated from carp fishes) was obtained from Fish Pathology Laboratory (ICAR-CIFA, Bhubaneswar) and grown in our laboratory using TSB medium (Himedia, Mumbai, IN) for 24 h at 25°C. Single colony isolated on TSB agar was inoculated in TSB broth. The culture of A. hydrophila was grown overnight, and 100 ìl of culture was inoculated into the fresh broth. Optical density was measured at regular intervals up to OD600=0.7 and culture was transferred to ice. The broth culture was centrifuged at 5000 × g for 15 min in a cooling (4°C) centrifuge (Eppendorf, Hamburg, DE). The pellets were washed twice with sterile phosphate buffer saline (PBS), pH 7.4 (Himedia, Mumbai, IN) and centrifuged. The final suspension was used to prepare 10 fold dilutions in 20 mL PBS having 106 cfu, 105 cfu, 104 cfu and 103 cfu mL . The dilutions were spread plated on TSB agar plates in duplicate for each dilution. The colony forming unit (cfu) count at optical density 600 nm was standardized with the same procedure before the experiment in three consecutive trials.

For the pathogenicity challenge test, we followed intra-peritoneal injection method using 1 ml Insulin syringe. After acclimatization in the aquarium for five days with proper aeration and feed, fish samples (33.32 ± 5.96 g) were separated into five different groups (12 fishes in each group). Four groups received 106 cfu, 105 cfu, 104 cfu and 103 cfu of bacterial culture per 100 g body weight respectively while the fifth group was injected with 100 ìl of PBS. The samples were observed up to 96 h, and dead fishes were removed for routine bacteriological examination. Final LD50 values were estimated according to the method of Miller and Tainter. The results were calculated as mean values from two independent experiments.

Pathogenicity experiment

A consecutive experiment was conducted in two different groups of fishes with 35 samples in each using the standardized LD50 dose to test the pathogenicity. Test group were injected with A. hydrophila intraperitoneally according to their body weight while the control group was injected with 100 ìl of PBS. The fishes were observed for any physiological and pathological symptoms at various time intervals. Three samples from test group were anesthetized using tricaine methanesulfonate (200 mg L) (Sigma-Aldrich, St. Louis, US) and blood was drawn from a caudal vein in 3.8% sodium citrate containing vials. The blood samples were centrifuged at 7000 × g for 10 min and plasma was collected in the fresh tube. Plasma was stored at -80°C until further use. The plasma samples were used to estimate the levels of superoxide dismutase (SOD) and cortisol using ELISA and colorimetric kits (Cayman, Ann Arbor, MI), respectively. The statistical analyses were performed by the standard curve plot of known concentration of SOD and cortisol. The data was analysed using independent sample t-test (á=0.05). Liver, head kidney and spleen tissues from each sample collected at different time intervals were also plated on TSB agar plates (amp+) to detect the presence of bacteria.

Identification of a. Hydrophila was also confirmed by pcr amplification and sequencing of 16s rrna. The genomic dna was isolated from the randomly selected clones (6 no) grown on tsb agar plates and 16s rrna was pcr amplified using the universal primer pairs (16s rrna f- agagtttgatcctggctcag; 16s rrna r-ggttaccttgttacgactt). The pcr reaction was performed in a total reaction volume of 50 ìl containing 1 × pcr buffer, 10 pm of each primer, 200 ìm each dntp, 50 ng template dna and 1u taq dna polymerase (invitrogen, waltham, usa). The reaction conditions were 94°c for 4 min; 35 cycles of denaturation at 94°c (30 s), annealing at 58°c (30 s) and extension at 72°c (90 s) followed by a final extension at 72°c for 10 min. The pcr products were sequenced using big dye terminator v3.1 cycle sequencing kit in abi3730 genetic analyser at scigenom laboratory, kochi. The nucleotide sequences so obtained were matched with ncbi genbank blastn toolkit using default settings.

Histopathology

Liver tissues collected at 48 h were processed for histological examination. Tissue samples were fixed in Bouin's Fixative and prepared for histological examination using recommended procedure. Tissues were further embedded in paraffin wax and sectioned at 5 ìm in a rotary microtome (Thermo Fisher Sci, Massachusetts, USA). The slides were stained with haematoxylin and eosin and observed under Leica DM2500 (Leica, Wetzlar, DE) to study any pathological changes/difference in tissues after acute bacterial infection.

Aeromonas hydrophila is the primary bacterial pathogen for many fish and aquatic organism, and thus, it is a great concern for the aquaculture industry. The fish under the present study is also a valued and potential species for aquaculture in the Himalayan cold water region. Therefore, the present study was aimed to examine the effect of Aeromonas hydrophilainfection on Golden Mahseer. Initially, we standardized the suitable temperature of 25°C for the growth of Aeromonas hydrophila in the laboratory condition as the ambient temperature in hill environment remains low (-2°C-25°C in different seasons). Many authors observed the growth of this bacterium within the range of 27°C-30°C which we could not see in our conditions. The fish-pathogenic Aeromonas are also reported to be psychrophilic in nature and was isolated from diseased rainbow trout that lives at very low temperature (5°C-20°C).

The level and dose of pathogenicity of the bacteria in several carp species were also found to be variableTherefore, the optimum dose of bacterial inoculation and pathogenicity was determined for the first time in this species and how this dose of bacteria interacts with the physiological mechanism was determined. Determination of LD50 in study organism before the experimental challenge is advantageous for successful experiment and induction of clinical signs and symptoms. Where higher dose may cause mortality, the suboptimal treatment may not enrich the desired genes. In the present study, an LD50 value of 1.74 × 105 cfu per 100 g of body weight was standardized for A. hydrophila in Golden Mahseer. LD50 value of A. hydrophila was previously determined for varies model and non-model fish species. The standardized LD50 value of A. hydrophila for in vivo experiments would serve as a baseline data for future immune response studies.

The intraperitoneal inoculation of A. hydrophila produced similar signs and symptoms, reported during the disease progression in other fish species. Clinical signs including weakness, slower movement, swimming closer to the surface, fin haemorrhages and red patches at the gut region were observed during 24-48 h p.i. Similar observations were documented in A. hydrophila infected catfish. The fishes showed noticeable signs of haemorrhages and abdominal swelling. Large external ulcerative lesions also developed in the injected area and around the anal orifice. Histological lesions were observed in liver and the spleen, and kidney tissues were found diffused due to acute infection. Sundus highlighted similar observations like structural fragmentation and necrosis of liver tissue and friable kidney. Aeromonas hydrophila may itself initiate pathological processes and induce inflammation. However, the role of extracellular products released during the infection may also play a significant role in causing degeneration and necrosis. Also, histological symptoms were not performed in other tissues as the liver was the primary site of infections due to Aeromonas infection. The infection was also very high in the liver in comparison to other tissues when plated on Agar plates at each time points of sample collection.

Dermal lesion with haemorrhagic foci may be correlated to bacterial haemorrhagic septicaemia mainly caused by Aeromonassp.The study findings are similar to Cipriano and Laith & Najiah who observed the dermal ulcerative lesions, focal haemorrhages and inflammation in chronic infection of Aeromonas sp. The histological findings of the study are similar to Laith and Najiah who reported that A.

hydrophila causes haemorrhagic sepsis, liver necrosis along with the accumulation of phagocytic cells at the site of necrosis. The results also agree with the finding of Afifi that various toxins and extracellular products such as hemolysin, protease and elastase released by A. hydrophila may cause severe hepatocyte necrosis. Similarly, necrosis of hepatocytes and the presence of non-hepatocytic cells, probably the phagocytic cells in liver histology also indicate the systemic bacterial infection in the liver which is also shown previously in various pathological studies.

Pathogenicity of Aeromonas was also determined using the stress related SOD and cortisol parameters. Both the SOD and cortisol were elevated at a particular time and scaled down during the longer periods when the fish overcame the pathogenicity. Superoxide dismutase (SOD) acts as cyclic enzymes that catalyze the dismutation of superoxide radicals. The amount of superoxide dismutase present in the cellular and extracellular environment is significant in the prevention of oxidative stress related diseases and can help to predict the defence mechanism of fish species. Production of reactive oxygen species (ROS) is also an important mechanism of the vertebrate immune system. They also possess various antioxidant enzymes to counteract the ROS and their adverse effects. In the present study, a significant increase in SOD activity was observed up to 24 h p.i. Wang studied the tissue-level activity and gene expression of two superoxide dismutase (MnSOD and Cu/Zn SOD) after A. hydrophila infection and found that the activity was highest in liver and kidney at 24 h p.i. However, a subsequent decrease in SOD activity suggests the failure of SOD to remove superoxide radicals. These results are similar to that of Reddy. The failure of SOD leads to accumulation of superoxides that ultimately may result in internal tissue necrosis.

The necrotic lesions observed in liver tissue at 48 h p.i. may be the combined effect of infection and ROS species. Although, cortisol inhibit the immune response by suppressing the inflammatory substances; the amount of cortisol may serve as a useful indicator of stress level during infection in fishes. The cortisol level was significantly ($p<0.05$) increased in blood plasma of Golden Mahseer after A. hydrophila challenge. The highest level was detected at 24 h p.i. which could be the combined effect of stress as well as bacterial infection. Increase in plasma cortisol level was reported in many piscine systems after bacterial infection. Small and Bilodeau also reported the similar findings on cortisol level for channel catfish (Ictalurus punctatus) after exposure to Edwardsiella ictaluri.

High cortisol level was observed in Atlantic cod after experimental infection with Aeromonas salmonicida . Increased cortisol level is commonly associated with stressful conditions of the fish principally during disease condition. Endotoxin may also increase the plasma cortisol level that may suppress the immune response of the fish. Hence, the increased level of plasma cortisol in Golden Mahseer may be associated with the stressful condition of fish as well as bacterial endotoxins. This is the first report on plasma cortisol level of Golden Mahseer in control and stress conditions. Handling during the bacterial and PBS inoculation as well as sampling may cause stress to the fishes, and its effect may not be neglected. However in the present study the cortisol level increased almost two folds in diseased fishes compare to control which is in agreement with the findings of Olsen and Haukenes and Barton who found the increased plasma cortisol level in diseased fishes.

CONCLUSION

In the present study, we determined the LD50 concentration (1.74 × 105 cfu per 100 g body weight) of A. hydrophila and examined various pathophysiological symptoms during the disease progression. The sign and symptoms of severe bacterial haemorrhage were observed, and A. hydrophila was confirmed to be the etiological agent by re-isolation of a bacterium using spread plate method followed by sequencing of 16s rRNA sequencing from infected fishes (test group). Two major stress indicators were also studied in test samples during the progression of infection. Superoxide dismutase activity and cortisol level were significantly high in test samples as compared to control. Severe liver necrosis and tissue damage were observed at 48 h p.i. Experimental conditions standardized in the present study will serve as a baseline data for further studies on Golden mahseer. Also, the knowledge of various symptoms during pathogenesis help to understand the course of pathogenesis and overall immune response of fishes which can be very beneficial to control and prevent the disease outbreaks in farmed fishes.

Aeromonas spp. are ubiquitous waterborne bacteria responsible for a wide spectrum of diseases among poikilothermic and homoiothermic organisms including humans (Austin and Austin 1987; Khardori and Fainstein 1988).

Two phenotypically distinct groups are well known within the genus Aeromonas: (i) psychrophilic and non motile Aeromonas salmonicida, which are relatively homologous phenotypically and (ii) mesophilic and motile aeromonads, which are heterologous. Within the latter group three phenospecies have been recognized, Aerom. hydrophila, Aerom. caviaeand Aerom. sobria (Popoff 1984). All these phenospecies are well known as fish pathogens (Toranzo et al. 1989; Candan et al. 1995; Koziñska 1996; Ogara et al. 1998; Austin and Austin 1999).

In recent years, the classification of the genus Aeromonas has undergone major changes. Extended DNA–DNA hybridization studies or other molecular techniques and improved phenotypic methods have resulted in the recognition of

14 well defined genomic species within the genus (Altwegg and Geiss 1989; Janda 1991; Martinez Murcia et al. 1992; Joseph and Carnahan 1994; Esteve et al. 1995b; Ali et al. 1996; Huys et al. 1997a,1997b). These species are as follows: Aerom. hydrophila, Aerom. bestiarum, Aerom. salmonicid (comprising non motile and psychrophilic as well as motile and mesophilic strains), Aerom. caviae, Aerom. media, Aerom. eucrenophila, Aerom. sobria, Aerom. veronii (with two biotypes sobria and veronii), Aerom. jandaei, Aerom. trota, Aerom. schubertii, Aerom. encheleia, Aerom. allosaccharophila and Aerom. popoffii.

A wide range of Aeromonas species have been found in association with diseased or healthy fish (Martinez Murcia et al. 1992; Sugita et al. 1994, 1995; Esteve et al. 1995a; Esteve et al.1995b; Huys et al. 1996), but only some of them have been documented as fish pathogens (Torres et al. 1993; Esteve et al. 1995a). Since the changes in Aeromonas classification, the importance of recently described species in fish pathology is mostly unknown. DNA–DNA hybridization techniques are used to accurately identify isolates to the genomospecies level, but are not available in routine ichthyopathology laboratories. For this reason, most of these laboratories identify motile Aeromonas as members of the 'Aerom. hydrophila' complex, which includes Aerom. hydrophila, Aerom. bestiarum and the motile biogroup Aerom. salmonicida; the 'Aerom. caviae' complex, which includes Aerom. caviae, Aerom. media and Aerom. eucrenophila; and the 'Aerom. sobria' complex, which includes Aerom. sobria, Aerom. veronii, Aerom. jandaei and Aerom. trota (Janda 2001).

Recently, 16S rDNA restriction fragment length polymorphism (RFLP) methods have been developed for the identification of all known Aeromonas species (Borrel et al. 1997; Figueras et al. 2000). Moreover, several schemes for the phenotypic identification of Aeromonasgenomospecies from clinical and environmental sources have been proposed (Altwegg et al. 1990; Abbot et al. 1992; Oakey et al. 1996; Brunner and Stolle 1997; Borrel et al. 1998). The identification of some of the genomic species mentioned above is still difficult because of some discrepancies between the DNA homology and phenotypic characters of various Aeromonas species.

There has been only a limited amount of investigation into the association of specific biochemical characters with specific Aeromonas species originating from fish (Austin et al. 1989; Esteve 1995; Noterdaeme et al. 1996; Carson et al. 2001; González et al. 2001). In the present study, numerous strains from common carp (Cyprinus carpio L.) were identified by 16S rDNA RFLP analysis and characterized phenotypically. The relationship between the genomospecies, phenotypic profile and pathogenicity for carp of these strains was evaluated.

Materials and methods

Bacterial strains

years. Bacteria were isolated by inoculating plates of R S agar (Thirty four strains of motile aeromonads isolated from healthy or diseased carp and two strains from tank water were used in this study. All strains were obtained from carp farms in Poland during the last 10Shotts and Rimler 1973h. Yellow colonies were then selected and identified to the genus) with fish skin and kidney or water samples and incubating the cultures at 28°C for 24 Aeromonas with a panel of tests according to Popoff (1984). All strains were Gram negative rods, positive for oxidase, catalase, motility, glucose oxidation/fermentation (O/F test), 0/129 resistance and growth in nutrient broth without but not with 6% NaCl. Additionally, one strain previously identified biochemically as non pigmented Chromobacterium violaceum like (Koziñska and Antychowicz 1996) was used.

Identification to the genomospecies level

All strains were identified by the 16S rDNA RFLP technique as previously described (Borrel et al. 1997; Figueras et al. 2000). In brief, the method consists of the digestion of the complete polymerase chain reaction amplified 16S rDNA gene using two endonucleases (AluI and MboI) simultaneously, which enables the identification of Aerom. hydrophila, Aerom. caviae, Aerom. media, Aerom. veronii, Aerom. jandaei, Aerom. eucrenophila, Aerom. sobria, Aerom. trota, Aerom. allosaccharophila and Aerom. schubertii. However, Aerom. bestiarum, Aerom. salmonicida, Aerom. popofii and Aerom. encheleia exhibit the same RFLP pattern after this digestion (Borrel et al. 1997). To identify Aerom. popofii and Aerom. encheleia, further digestion with the enzyme NarI followed by HaeIII and AluNI is necessary while for Aerom. bestiarum and Aerom. salmonicida NarI followed by PstI is required (Figueras et al. 2000).

Extended phenotypic studies

All strains were tested for 20 characteristics by the API 20E system, 19 characteristics by the API Zym system (bio Mérieux, MarcyL'Etoile, France) and 26 other phenotypic characters. Selected biochemical tests were carried out according to Abbot et al. (1992) and Carson et al. (2001) enabling discrimination of Aeromonas genomospecies. These tests included gas production from glucose, aesculin hydrolysis, salicin fermentation, gluconate oxidation, elastase production, H2S production from cysteine, dl lactate utilization, phenylalanine deaminase, acid production from dh at 100°C) and haemolysis activity (on trypticase soy broth supplemented with 5% (v/v) horse blood) Additionally, ornithine decarboxylase, lysine decarboxylase, arginine dihydrolase, citrate utilization (Simmons) and fermentation of mannose, cellobiose, glycerol, lactose, resistance to cephalothin and ampicillin.

Moreover, all strains were characterized for the fermentation of xylose, trehalose, maltose, fructose and raffinose, growth at 4, 37 and 42°C, agglutination in brain heart infusion (BHI) broth before and after boiling (1 d mannitol and ld for the remaining tests. h for API 20E, API Zym, haemolysis and autoagglutination or after 2–7 arabinose were performed using conventional tube tests to confirm the results obtained by the API 20E tests.

PATHOGENICITY ASSAYS

Common carp (C. carpio l mg weeks and fed commercial pellets (Carp K30; Aqua Pasze, Olsztyn, Poland) in order to adapt to laboratory conditions. All fish were anaesthetized with MS 222 (Sigma, St. Louis, MO, USA) (80 g average weight were obtained from a fish farm in Poland and certified as disease free. The fish were maintained in 100 l aquaria with aerated fresh water at 16°C for 2 L.) of 150 ml of a suspension containing 10) and then injected subcutaneously with 0·1 d were tested bacteriologically for the presence of injected bacterium in damaged fish tissues. d and the virulence level of the strains estimated and grouped as follows: more than six fish with disease symptoms and mortality of five to 10 fish (strongly virulent); more than six fish with disease symptoms and mortality of one to four fish (virulent); more than four fish with disease symptoms and without mortality (weakly virulent) and one to four fish with slight pathological signs and without mortality or neither pathological signs nor mortality noted (avirulent).

Recently dead fish and any survivors after 7 ml of sterile PBS as controls. The fish were then replaced into the aquaria and held under the same conditions as before injection. The morbidity and death of the fish were monitored daily for 7 bacterial cells in phosphate buffered saline (PBS). Ten fish were injected per strain and ten other fish were injected with 0·1

Emerging Aeromonas Species Infections and Their Significance in Public Health

Aeromonas species are ubiquitous bacteria in terrestrial and aquatic milieus. They are becoming renowned as enteric pathogens of serious public health concern as they acquire a number of virulence determinants that are linked with human diseases, such as gastroenteritis, soft-tissue, muscle infections, septicemia, and skin diseases. Proper sanitary procedures are essential in the prevention of the spread of Aeromonasinfections. Oral fluid electrolyte substitution is employed in the prevention of dehydration, and broad-spectrum antibiotics are used in severe Aeromonas outbreaks. This review presents an overview of emerging Aeromonas infections and proposes the need for actions necessary for establishing adequate prevention measures against the infections.

The aeromonads are Gram-negative, rod-shaped, facultative anaerobic, nonspore forming bacteria that are autochthonous and widely distributed in aquatic environments.

The genus is made up of psychrophiles and mesophiles from soil and aquatic environments and causes different kinds of diseases to many warm and cold-blooded animals. Several reconsiderations on the taxonomy and nomenclature of Aeromonasgenus have been carried out over the years. Although Aeromonas was initially positioned in the family Vibrionaceae, successive phylogenetic analyses point out that the genus Aeromonas is not closely related to vibrios resulting in the relocation of Aeromonas from the family Vibrionaceae to a new family, the Aeromonadaceae. The aeromonads and Enterobacteriaceae share many biochemical characteristics but are easily differentiated by oxidase test for which the aeromonads are positive. Generally, members of the genus are characteristically divided into three biochemically differentiated groups (Aeromonas hydrophila, Aeromonas caviae, and Aeromonas sobria), and these contain a number of genomospecies, and recently, new species have been added. Currently, the genus comprises of 17 DNA hybridization groups (HGs) or genomospecies and 14 phenospecies.

Aeromonas species are known as causative agents of a wide spectrum of diseases in man and animals. Some studies have shown that some motile Aeromonas species are becoming food and waterborne pathogens of increasing importance. They have been associated with several food-borne outbreaks and are progressively being isolated from patients with traveler's diarrhea. Presently, as a putatively emerging enteric pathogen, Aeromonas species have the inherent capability to grow in water distribution systems, especially in biofilms, where they may be resistant to chlorination. Also, A. hydrophila is listed in the Contaminant Candidate List, and Environmental Protection Agency Method 1605 has validated its detection and enumeration in drinking water system.

In 1968, von Graevenitz and Mensch reviewed 30 cases of Aeromonas infections or colonization, which created awareness on their recognition as human pathogens and suggested that some aeromonads may be associated with gastrointestinal diseases. Aeromonas species are commonly isolated from fecal sample of children under the age of five years, whereas their isolation from other body sites usually occurred in adult populations. Aeromonads are known to cause severe diarrheal disease of short duration or chronic loose stools in children, the elderly, or the immuno-compromised individuals, and they have been implicated in travelers' diarrhea. The bacteria also cause cellulitis or wound infections due of traumatic injury in aqueous environment. They also cause septicemia related to underlying diseases such as leukemia, cancer, cirrhosis, and various infections such as urinary tract infections, surgical wound infections, meningitis, peritonitis, and endocarditis. Some predisposing factors for Aeromonas infection include hepatic diseases, diabetes, hematologic malignancies, hepatobiliary, and renal diseases.

Environmental source of Aeromonas pathogen involved in gastrointestinal infection was first reported by Holmberg and coworkers. It was presented as an epidemiological study to backup the significance of untreated well water as a source of infection in patients with diarrheal disease. Phenotypic identification was relied upon before 1990s for the identification of Aeromonas, while several studies published thereafter identified isolates to hybridization groups. Hybridization groups containing virulence factors may be found in environmental and foods samples, but aeromonads will only cause gastroenteritis when their presence goes beyond an infective dose for a vulnerable host. In this paper, we present a general overview of Aeromonas species in the light of the increasing report of this group of bacteria as emerging pathogens and their effect on public health.

Occurrence of Aeromonas in Aquatic Environment

Aeromonas species have been found in chlorinated drinking water supplies in several countries, and typically of densities less than 10 cfu/mL in drinking water distribution systems. They occur in distribution system biofilms where they may be protected from disinfection, and multiple strains are frequently found in water sources.. In the Netherlands, a drinking water standard of 200 cfu/100 mL at 25°C has been established for Aeromonas.

Maalej and colleagues studied seasonal distribution of Aeromonas populations in urban effluent and natural seawater. They found 1.48 × 10 to 2.2 × 10 cfu/100 mL in the effluent, while seawater had counts lower than the detection limit to 7.9 × 10 cfu/100 mL, and the seasonal abundance of Aeromonas was inversely related to the seasonal density of fecal coliforms.

Bonadonna et al. studied the incidence of bacteria of anthropomorphic origin and those of autochthonous origin using model systems for prediction of public health risk to marine bathers. The resulting model used salinity, total coliforms, fecal coliforms, Escherichia coli, and location as analytical variables for presence of aeromonads. In the study, the presence of Escherichia coli and fecal coliforms were associated with lower Aeromonas counts, while prevalence of total coliform was associated with higher Aeromonas densities. Fecal coliforms and high salinity was associated with higher Aeromonas population. The complexity of the relationship between anthropomorphic and autochthonous bacteria confounds development of a predictive models for estimating public health risk of recreational exposure to marine waters.

Occurrence of Aeromonas in Animal

The isolation of Aeromonas species from diseased fish, turtles, alligators, snakes, and frogs was the first implication of Aeromonas as animal pathogens. A. salmonicida and A. hydrophila cause furunculosis, ulcerative disease, hemorrhagic disease, red sore disease, and septicemia in fish. Lehane and Rawlin investigated zoonoses acquired from fish and reported that aeromonads caused cellulitis, myositis, and septicemia as a result of injuries from handling fish, working in aquaculture, or keeping fish as pets. Also, Aeromonas species have long been recognized as pathogens of amphibians and reptiles.

A. hydrophila have been isolated from feces of normal horses (6.4%), pigs (9.6%), sheep (9.0%), and cows (21.1%). The total fecal carriage rate in animals is slightly higher than that in normal humans, which is <1 to 7% for most studies, although some studies report higher rates. Figura and Marri isolated Aeromonas species from the feces of domestic animals and found higher frequency of A. hydrophila than A. caviae. Stern et al. isolated aeromonads from 3 of 21 turkeys and 1 of 32 cows, but none was detected from 22 pigs and 24 sheep. Gray and Stickler reported findings of A. hydrophilapredominantly in cow feces and A. caviae in pig feces. Also, Hathcock et al. reported the isolation of Aeromonas species from feces, bedding, and drinking water of healthy cows and pigs.

Foods of animal origin including contaminated animals may play significant roles in the transmission of the aeromonads from food or animals to humans and animal feces appear to be the major source of contamination of foods. Aeromonads have recurrently been isolated from meat and edible organs of sheep and poultry, fish and seafood, raw milk, red meats, as well as pork and beef. A. hydrophila was also isolated from diseased cyprinid loach in Korea. The presence of aeromonads in food of animal origin may pose a public health problem for humans who come in contact with such animals.

Occurrence of Aeromonas in Human

Humans carry Aeromonas species in their gastrointestinal tract both in symptomatic and asymptomatic individuals. The rates of faecal carriage in persons in the absence of disease in developed countries range from 0% to 4%, while the isolation rate from persons with diarrheal illness ranges from 0.8 to 7.4%

Sinha et al. reported aeromonads in 6.5% of all patients in India, while in Hong Kong, Chan et al. documented aeromonads in 6.9% of adult patients with acute diarrhoea. Others reported isolation rates from symptomatic patients from 0.04% to 21%. In Limpopo Province of South Africa, Obi and Bessong reported the isolation of Aeromonas sp. from 13.3% HIV patients with chronic diarrhoea in rural communities. Immunocompromised people can also suffer from Aeromonas-associated chronic diarrhoea. In Saudi Arabia, Ibrahim and Colleagues reported two cases of chronic colitis from immunocompromised patients associated with A. hydrophila.

AEROMONAS INFECTIONS AND THEIR TREATMENT.

With advances in the identification and molecular taxonomy of Aeromonas spp., these organisms, which are widely distributed in the environment, are increasingly being recognised as human pathogens. Clinical infections include gastroenteritis, skin and soft tissue infections and bacteraemia. Antibiotic resistance poses a potential problem in the antimicrobial therapy of infections cased by Aeromonas spp. While most strains are susceptible to chloramphenicol, ciprofloxacin, co-trimoxazole and the aminoglycosides, the activity of amoxycillin/clavulanate and the acylureidopenicillins is inconsistent. Addition of a beta-lactamase inhibitor does not significantly enhance the activity of the acylureidopenicillins. Aztreonam and the carbapenems, imipenem and meropenem remain highly active. Although resistance to the first and second generation cephalosporins is variable, more than 90% of Aeromonas spp. are susceptible to the third generation agents. Of potential significance is the identification of chromosomally-encoded inducible beta-lactamases, associated with resistance to extended spectrum penicillins, cephalosporins, monobactams and carbapenems, in clinical isolates of Aeromonas spp. Two distinct enzymes are produced: the A1 enzyme, a serine beta-lactamase behaving as a group 1 cephalosporinase, and the A2 enzyme, a metallo beta-lactamase which hydrolyses a wide range of beta-lactam agents including the carbapenems. The clinical relevance of these enzymes in Aeromonas spp. is unclear.

Outbreak of Aeromonas hydrophila Wound Infections Associated with Mud Football

Members of the genus Aeromonas are facultative anaerobic, nonsporulating gram-negative bacilli that are ubiquitous inhabitants of fresh and brackish water. Aeromonas species have been found in a variety of aquatic environments, including lakes, rivers, streams, springs, rainwater, swimming pools, and seawater, and have also been isolated from tap water and soil. These species have been recognized as pathogens of fish, reptiles, and amphibians for many decades, but it is only recently that they have been recognized as significant human pathogens. In humans, infections caused by Aeromonas species generally result in either acute or chronic gastrointestinal illness, septicemia in immunosuppressed individuals, or water- or soil-associated traumatic wound infections.

Aeromonas wound infections are most commonly caused by Aeromonas hydrophila and have been reported after accidental puncture of the skin followed by exposure to contaminated water or soil [4–8]. These infections occur sporadically and infrequently, and they are more common in warmer climates. Wound infections caused by A. hydrophila often progress rapidly and may require surgical debridement or the amputation of limbs or digits. Fatal Aeromonas wound infections in healthy adults have also been reported. Treatment of Aeromonas wound infections is complicated by the fact that members of this genus are universally resistant to penicillin (the result of the presence of chromosomal â-lactamase), rendering standard empirical antibiotic treatment for common streptococcal or staphylococcal wound infections ineffective.

In this report, we describe an unusual outbreak of wound infections caused by A. hydrophila in individuals participating in a "mud football" competition in a small rural town in the southwest of Western Australia. An investigation was conducted to ensure that appropriate antibiotic therapy was administered to patients, to identify factors contributing to the outbreak, and to add to our knowledge of the clinical features of A. hydrophila wound infections.

Patients and Methods

Collie is a small rural town of <“8500 residents situated 200 km south of Perth, the capital of Western Australia. On Sunday, 17 February 2002, a total of 26 persons who had participated in a charity mud football competition in Collie on the previous day presented to the emergency department at Collie Hospital with infected scratches and pustules over their torsos and limbs. Most persons reported 20–30 lesions, with some reporting >100 lesions. One patient required removal of an infected thumbnail at the emergency department, and another required surgical debridement of an infected toe in the hospital the next day. Swab samples were obtained from lesions of 2 patients at the emergency department, and A. hydrophila was identified from cultures of these swabs 2 days later. A third swab specimen obtained from the patient requiring surgical debridement also grew A. hydrophila. Anecdotally, at least 16 mud football players, in addition to the 26 who presented at the emergency department, visited their medical practitioners with similar symptoms, but no further data were obtained from these individuals. Eleven adult and 4 youth teams consisting of a total of <“100 people were involved in the mud football competition. The games were played between 1:30 pm and 4:00 pm in the afternoon on a midsummer day in which the maximum temperature reached 26°C. Two football fields were used simultaneously for a round-robin competition, with each game consisting of two 15-min halves. A Rugby Union competition was played, which involved considerable physical contact, including players tackling and wrestling each other for the ball in the mud. Children who did not participate in the mud football competition were provided with their own mud pool.

Case series

All 26 patients (or their parents) who presented to the emergency department at Collie Hospital on Sunday, 17 February, were interviewed. A questionnaire was used for the interview that addressed the clinical features and exposure of the patients to mud and to river water. Other data collected included the estimated number and location of skin lesions, other presenting symptoms, preexisting medical conditions, and current systemic antibiotic treatment. Patients who had been prescribed systemic antibiotics before the identification of A. hydrophila were advised to contact their medical practitioner to ensure that they were taking the recommended antibiotics for treatment of Aeromonas infection. These patients were also followed-up to determine whether their antibiotic treatment was altered.

Environmental investigation

An inspection of the mud football fields, the adjacent Collie River, and the irrigation equipment was performed by the local environmental health officer. In addition, a water sample was obtained from the river near the inlet pipe for the irrigation pump and tested for temperature, pH, and bacterial pathogens.

Laboratory methods

Swab samples of skin lesions were plated onto horse blood agar plates. After overnight incubation at 35°C, oxidase-positive colonies were further identified with the API 20E biochemical identification system (BioMerieux). Antibiotic susceptibility testing of clinical isolates was completed using the NCCLS agar dilution method.

To test water samples for Aeromonas species, 100 mL of water was filtered through a 0.45-μm nitrocellulose membrane. This membrane was then placed on a horse blood agar plate containing ampicillin (5 mg/L). After incubation overnight at 37°C, oxidase-positive colonies were further identified with the API 20E biochemical identification system.

Results

Public health management of the infection outbreak.

After being notified of the outbreak of wound infections early, the Collie environmental health officer compiled a list of mud football participants so that they could be telephoned, assessed, and advised about the appropriate management of their lesions. The next morning, putative Aeromonas species were reported by the Collie microbiology laboratory and sent to the Public Health Reference Laboratory in Perth for speciation and antibiotic susceptibility testing. The Communicable Disease Control Branch in Perth was notified by the Reference Laboratory of the cultures positive for Aeromonasspecies and began to coordinate the response to this outbreak with the regional public health unit.

The primary objective of the public health response was to ensure that all patients with Aeromonas infection were identified and provided with appropriate treatment and that medical practitioners and the public were alerted to the possibility of Aeromonas wound infections due to mud football or exposure to river water. Letters and Aeromonas infection fact sheets were faxed to local medical practitioners advising them to review their treatment of any mud football participants they had seen with skin infections, particularly any antibiotic therapy that was administered. The treatment recommended for suspected Aeromonas skin infections was oral ciprofloxacin, oral trimethoprim-sulfamethoxazole, or intravenous ceftriaxone. Patients who had presented to the emergency department on Sunday were interviewed and were advised to contact their medical practitioner for reassessment of their antibiotic therapy. The Communicable Disease Control Branch also released a local media statement warning the public about the potential for serious infections after exposure to mud or untreated water supplies and e-mailed all local microbiology laboratories to advise them to be alert for the presence of Aeromonas species in wound isolates.

Antibiotic susceptibility.

The resistance profiles of the 3 clinical isolates of A. hydrophila were identical. These isolates were found to be resistant to amoxicillin, meropenem, oral cephalosporins (cefaclor and cephalexin), cephalothin, and colistin and were susceptible to norfloxacin, ciprofloxacin, gentamicin, tobramycin, amikacin, trimethoprim, ceftriaxone, ceftazidime, amoxicillin-clavulanate potassium, ticarcillin disodium-clavulanate potassium, aztreonam, cefepime, and nitrofurantoin.

The reported locations of lesions were the legs (77%), arms (58%), torso (35%), back (23%), chest (19%), buttocks (8%), feet (8%), head (4%), and face (4%). The emergency department physicians who treated these patients reported that only scratches and abrasions were infected (i.e., there were no infected lesions on intact skin) and that up to 50% of all scratches were infected in some patients (M. J. Birch and B. Saharay, personal communication).

Twenty-two players (85%) reported symptoms in addition to infected lesions, including rash (69%), malaise (46%), fever (35%), headache (35%), myalgia (31%), nausea (31%), rigors (8%), sore throat (4%), and earache (4%). Although rash was reported by a large number of patients, attending physicians did not substantiate this, suggesting that patients confused their numerous cuts and abrasions with the presence of a rash. None of the players who presented to the emergency department reported any immunocompromising illnesses.

Antibiotic therapy was prescribed for 23 (88%) of the 26 patients presenting to the Collie Hospital emergency department (table 1). In all of these patients, the empirically provided antibiotic therapy was unlikely to be effective against A. hydrophila infection. After reassessment of their clinical status, 15 (65%) of the 23 patients had their antibiotic regimen changed. The treatment in 8 patients was not changed, because their skin infections were resolving or had resolved. Of the 3 patients who were not initially prescribed antibiotics, 1 (patient 18) was provided ciprofloxacin after reassessment by her doctor. In addition, the patient who required toe surgery (patient 15) was administered intravenous ceftriaxone therapy after initially being treated with flucloxicillin by emergency department physicians.

Environmental investigation

The environmental health officer reported that the mud football fields were prepared by plowing them to a depth of 500–600 mm and then irrigating them with water from the adjacent Collie River with sprinklers. At this time, the Collie River was low and had pooled as a result of low rainfall levels (>30% below average; Bureau of Meteorology, Perth, Western Australia) during the previous 12 months. Water was pumped onto the fields with an irrigation pump and PVC pipes that formed part of an orchard irrigation system >25 years old. The fields were irrigated each evening for 1 month before the mud football competition, with the amount of watering increased a few days before the event to saturate the fields.

The fields were originally used to grow fruit trees, but these trees had been removed, and the fields had been fallow for 2 years. The soil consisted of pea gravel and contained a stubble that was a mixture of wild oats, wild turnips and radishes, field grasses, and weeds. A. hydrophila was cultured from the sample of river water obtained from near the irrigation pipe inlet at the time of the outbreak. The pH of the river water was 7.5, and the water temperature near the irrigation pipe was 23°C. The surface water temperature of the parts of the river that received more sun exposure was <“30°C.

Aeromonas infections

The genus Aeromonas consists of gram-negative rods widely distributed in freshwater, estuarine, and marine environments. Aeromonas species grow at a range of temperatures, although they are isolated with increasing frequency during warmer months (May through October in the Northern hemisphere). Aeromonas species cause a wide spectrum of disease syndromes among warm- and cold-blooded animals, including fish, reptiles, amphibians, mammals, and humans.

The genus Aeromonas was re-categorized from the family Vibrionaceae to the family Aeromonadaceae in the mid-1980s, when phylogenetic evidence from molecular studies became available to support this distinction.

The genus Aeromonas has been divided into two major groups :

Motile, mesophilic species, including eight that can cause disease in humans.

Non-motile, psychrophilic species that generally cause disease only in fish.

Aeromonas species are oxidase positive and ferment glucose. The organisms grow at a range of temperatures from 0 to 42°C.

Pathogenicity and Pathogenesis

The role of Aeromonas as a gastrointestinal pathogen is uncertain. Stool isolation rates among individuals with diarrhea are variable, ranging from <1 to >60 percent. In addition, Aeromonas is a common isolate from asymptomatic individuals. There has been only one major outbreak in which Aeromonas has been implicated as the etiologic agent of disease . Efforts to induce illness in volunteers with selected Aeromonas strains were unsuccessful in one trial, although selection of strains may not have been optimal.

Nonetheless, it is likely that certain strains of Aeromonas cause diarrheal disease. One large study identified Aeromonas spp as a cause of traveler's diarrhea in 18 (2 percent) of 863 patients in Spain. In the Global Enteric Multicenter Study (GEMS), which compared stool cultures among 12,110 children with diarrhea and 17,291 matched control children at seven global sites, Aeromonas was associated with diarrhea only in Pakistan and Bangladesh, where Aeromonas-positive culture rates ranged from 19 percent in the 0- to 11-month age group to 29 percent in the 24- to 59-month age group. However, Aeromonas was isolated as the sole pathogen in less than 5 percent of cases at those sites, with Shigella as the most common co-isolate. In contrast, in Africa and India, rates of isolation never exceeded 1 percent. Factors responsible for these regional differences in incidence remain to be determined, and the potential role of coinfecting pathogens, such as Shigella, in disease occurrence warrants further study.

At a patient level, it is not possible to predict whether a specific Aeromonas strain is responsible for diarrheal illness. Possible virulence factors of Aeromonas species include toxins (cytotoxic and cytotonic), proteases, hemolysins, lipases, adhesins, agglutinins, pili, enterotoxins, various enzymes, and outer membrane arrays, such as an S-layer and capsule. Other factors that may contribute to virulence include VacB , enolase , and the presence of a Type VI secretion system. In one study, whole genome sequencing and comparative genomics were used to characterize a virulent subtype of Aeromonas hydrophila isolated from aquatic wounds that has genes for toxin production, toxin secretion, and bacterial motility . It is uncertain how many aeromonads contain these putative virulence factors and how these factors contribute to risk of illness.

Mesophilic aeromonads have a global distribution and have been isolated from a variety of aquatic environments, including:

- Fresh water
- Estuarine (brackish) water
- Surface water, especially recreational
- Drinking water, including treated, well, and bottled
- Polluted waters
- Waste water effluent sludge.

Aeromonads are not generally considered marine organisms, but can be found in marine systems that interface with fresh waters and can survive at all but the most extreme salt concentrations . Usually they are not part of the groundwater bacterial population, which is generally poor in nutrients.

In nutrient-rich waters, Aeromonas species can grow to large numbers and generally peak in the warmer temperatures of the summer months in temperate freshwater lakes and chlorinated drinking water. Aeromonas species appear to tolerate polluted environments, including chemical pollution, although they are not considered to be of fecal origin . The organism has also been isolated from retail produce sources and meat products .

Contact with any fresh or brackish water body is the most common source of human infection. The risk of infection can be reduced by caution in the setting of natural water sources (lakes, rivers, streams, ponds, bays), including minimizing the risk of traumatic wounds and avoiding oral ingestion, particularly during warmer summer months.

Underlying host factors may also influence the likelihood of infection. In the Global Enteric Multicenter Study (GEMS) study, breastfeeding was associated with a lower likelihood of Aeromonas infection, but only in patients 24 months of age or younger . In a study from Bangladesh, detection of Aeromonas in diarrheal samples was associated with severe acute malnutrition (odds ratio = 25.7). Hematologic malignancies and cirrhosis are associated with increased risk for infection.

Diarrheal disease is the most common manifestation of Aeromonas infection. The organism has also been associated with a variety of extraintestinal presentations.

Diarrhea — Aeromonas spp are associated with a range of diarrheal presentations including:

- Acute, secretory diarrhea, often accompanied by vomiting
- Acute, dysenteric diarrhea with blood and mucus
- Chronic diarrhea, lasting more than 10 days
- Choleric diarrhea with "rice-water" stools
- Traveler's diarrhea (probably the most commonly recognized presentation in the United States)
- There have been two reports of A. hydrophila enterocolitis associated with the hemolytic uremic syndrome (HUS).

Wound infections — Aeromonas can cause mild to severe wound infections. Infection typically occurs on the extremities following traumatic aquatic injury. Such wound infections affect men three times more commonly than women. The most typical presentation is cellulitis , although myonecrosis (with and without gas production), rhabdomyolysis, and lesions mimicking ecthyma gangrenosum have been reported. A case of nearly fatal necrotizing fasciitis from a traumatic leg wound incurred from contact with a fresh water river highlights the virulence potential of aeromonads to cause serious disease.

A. hydrophila, Aeromonas veronii, and Aeromonas schubertii are the species most commonly isolated from wound infections. Aeromonas was a common wound isolate among tsunami victims in southeast Asia in 2004, and elevated numbers of Aeromonas spp were recorded in floodwater samples in New Orleans following Hurricane Katrina in 2005. Additionally, several reports have documented aeromonad necrotizing fasciitis associated with species other than A. hydrophila, such as A. veronii biovar sobria, A. schubertii, and Aeromonas caviae. A former arabinose-negative biovar of A. hydrophila is now known as Aeromonas dhakensis and has a high association with more serious human infections, especially in Taiwan, Malaysia, and Australia .

Serious wound infections and sepsis have also been reported following the medicinal use of leeches [58-60]. Aeromonads reside in the gut of the leech Hirudo medicinalis, where they assist in the enzymatic digestion of the blood ingested by the leech [58,59,61,62]. Patients undergoing leech therapy often receive systemic chemoprophylaxis with ciprofloxacin to prevent such infection. Emerging reports of ciprofloxacin-resistant strains of Aeromonas isolated from leeches may limit the utility of this practice.

Bacteremia — Sepsis with Aeromonas species is strongly associated with infection with A. veronii biovar sobria. These patients present with the classic signs and symptoms of gram-negative sepsis and may have gastrointestinal symptoms, including abdominal pain, nausea, vomiting, and diarrhea.

Sepsis tends to occur in older patients with hematologic malignancy , serious hepatobiliary disease, other immunocompromising conditions, or traumatic injuries. One case report described recurrent Aeromonas bacteremia over two years in an elderly man who had repeated exposure through contaminated well water . Pediatric case reports of sepsis due to A. hydrophila have been reported, including one case of a patient with diarrhea and pneumonia and one case of acute renal failure. Cases have also been rarely reported among pregnant women; in 2011, three cases of bacteremia with Aeromonas spp were identified in pregnant women at the Thailand-Myanmar Border.

It is not always possible to identify the source of the organism in cases of sepsis; in such cases, it is reasonable to surmise that it was acquired from the gastrointestinal tract.

Miscellaneous extraintestinal sites — Aeromonas spp have been implicated in cases of ocular infections, osteomyelitis, meningitis, respiratory infections following "near drowning," pelvic abscesses, otitis, cystitis, endocarditis, peritonitis, cholecystitis, and joint infections. Necrotizing fasciitis and folliculitis due to A. hydrophila strain have also been reported.

Aeromonads are not routinely identified in most microbiology laboratories as part of the normal protocol for isolating stool pathogens. For cases in which Aeromonas is suspected clinically, the laboratory should be advised to look for the organism. It is readily identified in routine wound or blood cultures. Automated genetic identification systems can identify most true Aeromonas isolates to the level of A. hydrophila group or A. hydrophila/A. caviae. However, these identifications are often incomplete or erroneous due to insufficient discriminatory markers to detect interspecies differences. A study evaluating the ability of six commercial systems to identify clinical Aeromonas isolates noted that accuracy of these systems was limited by outdated databases and taxonomy, weak algorithms, and the need for impractical tests . Certain algorithms, however, performed well, such as the Aerokey II , which correctly identified 95.5 percent of 87 isolates to the species level. This dichotomous algorithm can differentiate the emerging virulent species known as A. dhakensis, which has been linked with severe infection, including septicemia.

Hemolysis is variable on blood agar media; most species display beta hemolysis. Although aeromonads grow on nearly all enteric media, they often are overlooked on MacConkey agar because A. caviae is lactose-positive just like Escherichia coli. Ampicillin-containing medium should not be used to suppress normal enteric flora, since a substantial portion of A. caviae isolates and all Aeromonas trota isolates are sensitive to ampicillin and therefore will not grow on such a medium. This is especially important in light of reports implicating A. trota in pancreatic abscess and septic shock with cirrhosis, and A. caviae in cystitis.

Aeromonas spp are oxidase-positive, polar flagellated, glucose-fermenting, facultatively anaerobic, gram-negative rods that are resistant to the vibriostatic agent O/129 and unable to grow in 6.5 percent NaCl. The presumptive identification of an isolate involves the initial separation from other oxidase-positive genera such as Vibrio and Plesiomonas to avoid misidentification. This can be accomplished with simple tests such as O/129 susceptibility, tolerance to various NaCl broth concentrations, and the ability to ferment inositol.

Antimicrobial resistance markers and susceptibility studies should be determined by either the standard agar dilution method or by Kirby-Bauerdisk diffusion method using the 2010 CLSI Standard M45-A2 for Aeromonas species . Some automated MIC systems, such as BioMerieux Vitek, Inc. may not be reliable for detection of beta-lactam resistance .

THERAPY

Antimicrobial susceptibility — Clinical studies have demonstrated differences in antimicrobial susceptibility between species, highlighting the importance of both species identification and susceptibility testing for all isolates, particularly in the setting of serious infection. With the caveat that there may be substantial differences in regional resistance patterns, most Aeromonas strains are resistant to penicillin, ampicillin, carbenicillin, and ticarcillin; most are susceptible to trimethoprim-sulfamethoxazole (TMP-SMX), fluoroquinolones, second- and third-generation cephalosporins, aminoglycosides, carbapenems, chloramphenicol, and tetracyclines. There have been several reports of infection with fluoroquinolone-resistant A. hydrophila strains following leech therapy.

Most Aeromonas species produce an inducible chromosomal beta-lactamase, which may not be detected by rapid commercial susceptibility systems . Aeromonads produce beta-lactamases from three different classes: a class C cephalosporinase, a class D penicillinase, and a class B metallo-beta-lactamase (MBL) of the "CphA" type . Two other MBLs (VIM and IMP) in strains of A. hydrophila and A. caviae have also been detected, encoded on an integron and a plasmid, respectively. One reported the emergence of multiple Aeromonas spp. displaying CphA-mediated carbapenem resistance in Australia; the predominant species found was A. dhakensis, previously known to be the cause of more serious human infections.

Regional resistance patterns include the following:

- In Bangladesh, Aeromonas spp have the highest reported level of multidrug resistance (MDR) among enteric pathogens, at 82 percent [99]. This includes resistance to fluroquinolones (73 percent of isolates), macrolides (99.8 percent), tetracycline (72 percent), and TMP-SMX (85percent).
- In Spain, resistance to nalidixic acid has been reported; a surveillance study of 43 strains reported that 26 percent of A. caviae, 20 percent of A. hydrophila, and 88 percent of A. veronii biotype sobria were resistant to nalidixic acid. Although still susceptible to ciprofloxacin, these strains had a mutation in the A subunit of DNA gyrase and could easily develop a second mutation resulting in resistance to ciprofloxacin .
- In China, colistin resistance (mcr-3) has been described.

Clinical approach — Most cases of Aeromonas-associated diarrhea are self-limited and can be managed with supportive therapy, including oral and intravenous rehydration. Based on anecdotal data, antibiotics may be of value in patients with severe diarrhea and/or a history of immunosuppression . Antibiotic therapy is also indicated in the setting of wound infection and bacteremia.

Given emerging patterns of antimicrobial resistance, antimicrobial susceptibility testing of isolates is essential to guide antibiotic selection. Pending species identification and susceptibility testing, initial empiric therapy of suspected Aeromonas spp infections with a fluoroquinolone, third-generation cephalosporin, and/or TMP-SMX would provide reasonable antimicrobial coverage. For infections acquired in areas with high levels of resistance, such as Bangladesh, therapy with a third-generation cephalosporin and/or an aminoglycoside should be considered. Once susceptibility testing results are available, the antibiotic regimen can be tailored appropriately to a single agent.

Guidelines from the Infectious Diseases Society of America suggest a combination of doxycycline plus either ciprofloxacin or ceftriaxone for treatment of necrotizing skin infections caused by Aeromonas spp. However, there are no data that indicate this combination therapy offers a benefit over monotherapy.

Nevertheless, certain agents should be avoided; therapy with ampicillin or first-generation cephalosporins is not appropriate. All species of clinical aeromonads are resistant to ampicillin except for A. trota and sometimes A. caviae. A. veronii biovar sobria (formerly Aeromonas sobria) is uniformly resistant to first-generation cephalosporins, but in vitro testing suggests that it is susceptible to third- and fourth-generation cephalosporins.

There are no clinical trial data to guide the duration of therapy; therefore, treatment should be guided by clinical response. Reasonable courses of therapy include three days of therapy for treatment of diarrhea, 7 to 10 days of therapy for treatment of wound infections, and two weeks of therapy for treatment of bacteremia. The course of therapy may need to be adjusted depending on individual circumstances, including immunosuppression or other underlying conditions.

12

LUMPY JAW DISEASE

Lumpy jaw usually affects the bony tissues of the head rather than the soft tissues (as opposed to Wooden Tongue). It is caused by infection with Actinomyces bovis, which is part of the normal bacterial flora of the upper digestive tract. The bacteria usually invade the skin through a wound or minor trauma caused by sticks, straw or barley awns, or when deciduous teeth are being shed.

The disease is chronic, with the clinical symptoms developing slowly. The initial swelling of the jaw bone/bones is usually painless and the animals exhibit few clinical symptoms. Once the disease has advanced, the lesions may affect the teeth and the position of the jaws, leading to inappetence and emaciation. Advanced lesions may show discharging sinuses with yellowish pus. The disease is sporadic in the UK but the incidence varies in different parts of the world.

Note: The symptoms can be similar to that of other diseases (including Foot and Mouth) so it is important to examine any animals with mouth symptoms (excess salivation, difficulty in chewing, teeth problems, malaligned jaw) early.

Actinomycosis, commonly called 'Lumpy Jaw', is caused by the bacteria Actinomyces bovis, which is a normal inhabitant of the bovine mouth. Actinomycosis is a chronic bacterial disease and is more common in cattle than in goats and sheep. The bacteria enter through cuts and abrasions (i.e. due to teeth eruptions or coarse feed) and migrate to the bone, leading to osteomyeltitis (inflammation and infection of the bony tissues). The mandible is affected more commonly than the maxilla. The defining feature of actinomycosis is the presence of a non-painful swelling under the jaw. This swelling can rupture and drain pus-type, smelly fluid which contaminates the environment.

CONTROL AND PREVENTION OF LUMPY JAW

Rough grazing

Avoiding rough grazing, soil-contaminated silage and "weedy" hay/haylage or straw for cattle feed usually prevents trauma to mucosal membranes in the mouth.

In the case of an isolated incident in an individual animal, the best approach to prevention and control is to isolate and humanely slaughter the affected animal as soon as economically feasible or when the animal starts exhibiting clinical symptoms of pain and inappetence.

In an outbreak, it is important to identify the causative factors of the predisposing lacerations in the mouths of the affected animals. This is likely to be a feed source that, once identified, should be removed. Again, humane slaughter of affected animals is necessary at the appearance of clinical symptoms.

As the causative bacteria do not survive long in the environment and are present in the mouths of healthy cattle, the removal of the predisposing feed source is adequate for controlling the disease within a herd. If the problem persists, it is important to look for further sources of predisposing factors and eliminate them.

Treating Lumpy Jaw

Veterinary attention should be sought for cases, and it is important to differentiate between other causes of facial swelling, such as localised abscesses or Actinobacillus (Wooden tongue) infection as they carry a better chance of recovery. Penicillin treatment can be used to treat less advanced cases. Due to the chronic and invasive nature of the lesions, most treatment results fail, and lesions may recur soon after. Humane slaughter is recommended if there is no response to treatment or if cases are losing excessive weight. When treatment is advocated on animal welfare grounds for temporary improvement (recently calved suckler cows, cows in late gestation), a vet should always be involved in the treatment of the condition.

Lumpy Jaw and Welfare

Any animal showing signs of teeth lesions, malalignment of the jaws, difficulty to chew and/or inappetence appear should receive veterinary attention and/or arrange for humane slaughter without delay.

Removal of predisposing feed or grazing is also important from the animal welfare point of view, in order to prevent further suffering.

Good Practice based on Current Knowledge

When dealing with individual animals with lumpy jaw or with outbreaks of the disease, it is good practice to:

Examine early any animals with "mouth" symptoms (excess salivation, difficulty in mastication, teeth problems etc.) and contact a vet if lumpy jaw is suspected

Humanely slaughter affected animals if the condition is advanced

If an affected animal is not slaughtered immediately for welfare reasons (suckler cow with a small calf, cow in late gestation etc.), its clinical condition should be assessed regularly by a veterinarian and treatment should be initiated if the animal is judged to be suffering from the symptoms

Always attempt to identify the predisposing factor (feed or grazing) and eliminate it immediately, even when only one animal is affected

Clinical signs

Lumpy jaw can occur in bony and soft tissues, but is predominantly seen in the bones of the upper and lower jaw.

The first symptom noted is usually swelling, as the bone becomes enlarged and honeycombed, and full of pus. In most cases, but not always, the swelling will break out through the skin, and the discharge will be very thick and sticky.

Lumpy jaw is usually progressive. As the bony swellings continue to enlarge, gross disfiguration of the head can occur, much body condition will be lost and death may result.

Lumpy jaw is a disease that produces permanent hard swellings on the jaw bones of cattle. Once an infection is established, bacterial by-products begin breaking down the bone. In response to the bone infection, the body tries to repair itself by creating new bone. This process creates honeycombed bone structures with tiny abscesses filled with pus. As the disease progresses, large cavities may open up, draining grainy pus. Because of the nature and location of the disease cattle may have difficulty eating and drinking. The animal often begins losing condition, and with time death may result.

How it is spread

The organism generally enters the body through cuts and abrasions in the mouth. The eruption of teeth is thought to play an important role. Altering grazing management to try to reduce exposure of cattle to coarse or prickly feed helps reduce the prevalence of these conditions.

Monitoring and action

In most cases, a person can diagnose this disease by closely inspecting the animal. The laboratory confirms the diagnosis by examining microscopic smears or culturing the organism.

Control

Cattle with actino lesions that are large and discharging should be destroyed on the property, not sent to saleyards or meatworks.

Treatment

The treatment of lumpy jaw is similar to the treatment for wooden tongue, but often ineffective. If the disease is detected early, the animal can be sold for slaughter while still in good condition, and while the lesion is not broken at the skin and discharging. Early detection and salvage slaughter of an actino animal prevents the development of an advanced case, which would constitute an animal welfare offence by the animal's owner or property manager under the Animal Care and Protection Act 2001.

Many affected animals are culled as prognosis is often poor and many animals never fully recover. If culling is not preferred, treatments include iodides and antibiotics (penicillin and streptomycin). Additionally, the affected portion of bone and any associated teeth in the surrounding area can be removed.

Actinomyces bovis

Actinomyces bovis is a branching, Gram-positive, rod-shaped bacterium of the genus Actinomyces. It is the causative agent of lumpy jaw (actinomycosis) in cattle, and occasionally causes infections in humans. A. bovis normally populates the gastrointestinal tract of healthy ruminants, but is opportunistic in nature and will move into tissues through ulcerations or abrasions of the mucosa to cause disease. Disease occurs when there is physical damage to the tissue of the mouth, allowing the bacteria to colonize the deep tissue and bone, typically affecting the mandible and maxilla.

Actinomycosis is pathognomonic for abscesses containing "sulfur" granules, and appears as basophilic colonies with club-shaped reaction products on a histological preparation. Lumpy jaw is commonly treated with broad-spectrum antibiotics with varying success, and can be a major economic loss for producers in countries where it is endemic. Because this organism is zoonotic, it is a human health concern and can cause granulomas, abscesses, skin lesions, and bronchopneumonia.

A. bovis was first described in 1877 by Bollinger and Harz, when they isolated the causative agent of lumpy jaw from a cow. Two theories for the source of infection have since been developed. The first is called the exogenous theory and was developed by Bostroem in 1891, when he suggested infection is caused by a foreign body, such as hard hay or grass awns. This theory has mostly been supported by clinical observations, and has been hard to demonstrate in a laboratory setting. The second is called the endogenous theory, which suggests that A. bovis is naturally found in the mouths of healthy animals. This theory has been supported by several researchers over the years as they have been able to isolate multiple strains of Actinomyces. In 1905, Wright described isolation techniques and colony morphology in different media. A. bovis has been isolated from tonsillar crypts, infected teeth, and dental scum of normal mouths from human and animal patients. Several researchers were then able to conclude that the bacteria are part of the normal flora of the mouth, making the mouth a source of disease. The number and virulence of bacterial particles involved in the infection have been suggested to contribute to whether or not disease occurs. Several researchers have tried to establish the infection in live animals with isolated bacterial cultures, but the results have been mixed, with some seeing development of disease and some seeing no change. In 1950, the lifecycle of A. bovis was described.

Morphology

A. bovis is a branching, rod-shaped bacterium. In its diploid phase, it is Gram-positive; during the haploid phase, it is club-shaped and is Gram-negative. A. bovis is catalase-negative, which can differentiate it from other anaerobic diptheroids. A. bovis microcolonies on brain heart infusion agar can be distinguished from the closely related organisms Actinomyces israelii and Actinomyces naeslundii. The ability for A. bovis to hydrolyze starch is another way to distinguish it from A. israelli and A. naeslundii.

Lifecycle

A. bovis goes through a two-phase cell cycle, a haploid phase and a diploid phase. The lifecycle begins with the haploid phase with the germination of a spore. The spore buds out into club shapes and branches as it undergoes nuclear replication in preparation for conjugation. Two specialized haploid cells then conjugate to become the initial cell of the diploid phase. This initial cell germinates and branches as it undergoes nuclear replication. A spore is then formed from a diploid phase branch, and the spore undergoes nuclear reduction, after which the spore buds off and completes the lifecycle.

ROLE IN DISEASE

Clinical presentation and diagnosis

Lumpy jaw (actinomycosis) in cattle can present itself in two main ways. One is as soft-tissue abscesses in the mouth and on the tongue. The other explains its common name, lumpy jaw, as A. bovis bacteria infect the mandibular bone, causing osteomyelitis and the formation of periosteal new bone, giving it the classical lumpy jaw appearance. It is caused by sharp objects, such as coarse feed materials or sticks, causing penetrative lesions in the mouth when ingested or by complication of periodontitis. Deeper lesions are required to cause infections in the bone, while shallower lesions are more likely to cause mouth abscesses. The bony lumps are usually not painful and can develop over weeks to months, causing facial distortion, dyspnea, and occasionally loose teeth. The abscesses within the bony lumps or soft tissue commonly consist of purulent granulation tissue, often gritty in texture, surrounded by a fibrous capsule.

To diagnose lumpy jaw, the fluids exuding from the bony lump or other abscesses are sampled or aspirated if the infection has not broken through the skin. Tissue samples from the lesion can be examined by histopathology after being embedded in paraffin wax, cut into 4- to 5-mm sections, and stained with hemolysin and eosin stains. The clubs could also be stained by carbol fuchsin with decolourisation by weak hydrochloric acid and counter-staining with methylene blue, Van Gieson's stain, or aniline gentian violet followed by treatment with alcohol. A. bovis can be identified histologically by its basophilic colonies and club-shaped reaction product visible in neutrophilic granulomatous lesions.

The club-shapes, also called Spendore-Hoeppli proteins, seen on histological slides are the result of the bacteria being encased in a calcium-phosphate protein complex, as the host attempts to wall off the infection. Another diagnostic feature is the presence of "sulfur" granules, which is pathognomonic for actinomycosis. These granules are small, yellow clumps of exudate that are visible to the naked eye with a similar appearance to grains of sulfur, and are mainly composed of bacterial filaments and host calcium phosphate.

Treatment

The efficacy of the clinical treatment for A. bovis lesions is not always successful, but conventional treatment plans involving combinations of broad-spectrum antibiotics and concurrent administration of iodide solutions have demonstrated success in reducing the inflammation and spread of the lesion. Initial treatment involves surgical debridement of purulent and soft-tissue lesions followed by daily cleaning and flushing of the infection site with saline and povidone-iodine solution, respectively, for about 1 week. Treatment success has been noted individually for both potassium iodide oral solution, and for intravenous sodium iodide, as well as when used conjunctively.

However, frequent treatment with iodide may lead to toxicity, and monitoring for clinical signs of iodine toxicity should be included throughout the course of treatment. Consequently, administration of iodide should be either reduced or discontinued if such signs of toxicity are observed. Antibiotic combinations of intramuscular penicillin and streptomycin are commonly used for 5–7 days, and have been widely associated with treatment success. Additional antibiotics such as oxytetracycline, florfenicol, and cloxacillin have also demonstrated sensitivity.

13

VIRAL DISEASE: INFECTIOUS SALMON ANEMIA

Infectious salmon anemia (ISA) is a viral disease of Atlantic salmon (Salmo salar) caused by Salmon isavirus. It affects fish farms in Canada, Norway, Scotland and Chile, causing severe losses to infected farms. ISA has been a World Organisation for Animal Health notifiable disease since 1990. In the EU, it is classified as a non-exotic disease, and is monitored by the European Community Reference Laboratory for Fish Diseases.

ISA is caused by the infectious salmon anemia virus (ISAV). ISAV, a segmented RNA virus that is the only species in the genus "Isavirus", which is in the family Orthomyxoviridae, and therefore related to the influenza viruses. The genome encodes at least 10 proteins.

There are several distinct strains of the virus. The most common are a European strain and a North American strain.

PATHOLOGY

ISA virus causes severe anemia in infected fish. Unlike the mature red blood cells of mammals, the mature red blood cells of fish contain DNA, and can become infected by viruses. The fish develop pale gills, and may swim close to the water surface, gulping for air. However, the fish may show no external signs of illness and maintain a normal appetite, until suddenly dying. The disease can progress slowly throughout an infected farm and, in the worst cases, death rates may approach 100%. Post-mortem examination of the fish has shown a wide range of causes of death. The liver and spleen may be swollen, congested or partially already dead. The circulatory system may stop working, and the blood may be contaminated with dead blood cells. Red blood cells still present burst easily, and the numbers of immature and damaged blood cells are increased.

Infectious salmon anemia appears to be most like influenza viruses. Its mode of transfer and the natural reservoirs of infectious salmon anemia virus are not fully understood. Apart from Atlantic salmon, both sea-run Brown trout (Salmo trutta) and Rainbow trout (Onchorhyncus mykiss) can be infected, but do not become sick, so it is thought possible that these species may act as notable carriers and reservoirs of the virus.

Research shows that while several species of Pacific salmon can be carriers of the virus, even highly pathogenic strains, Pacific salmon currently show high relative resistance and no ISAV-related symptoms. However, the potential for ISAV adaptation to Pacific salmon exists.

Evolution

There are several distinct strains of the virus, some are pathogenic and some are not. The most common are a European strain and a North American strain.

Genetic research into the ISA virus shows that the European and North American strains of the virus diverged from each other sometime around 1900. This research points out that starting in 1879 when rainbow trout were first brought to Europe from North America, there were many transfers of fish across the Atlantic ocean which may have carried the ISA virus. Some species were introduced to Europe from North America, and some species were introduced to North America from Europe. Given that the virus did not evolve into two separate strains until around 1900, and given how many transfers of fish there were, according to this research, "At present it is therefore not possible to suggest a geographical origin of the ISA virus."

Another study suggests that this virus was introduced into Norway between 1932 and 1959 and that the original strain was the European subtype found in North America. The strains found in Chile were transmitted from Norway between 1995 and 2007.

In Norway, cases of ISA have occasionally been reported in fresh water farms but generally in hatcheries which use part sea water. The overwhelming majority of cases occur in farmed fish in sea water. The virus has been detected in wild fish but cases of clinical disease have only been reported in farmed fish. The disease has also been detected in the UK, Canada and Chile.

Diagnosis

Cases of clinical disease may be characterised by severe anaemia, ascites (accumulation of fluid in the body cavity), haemorrhage in internal organs and darkening of the liver. Fish also tend to appear lethargic with pale gills and blood spots in the eyes.

The causative agent of ISA is an orthomyxovirus. The presence of the virus is confirmed by isolation and identification techniques, indirect fluorescent antibody test (IFAT) and reverse transcriptase polymerase chain reaction (RT-PCR).

Control

The virus can be transmitted through water, but the highest risk factors for spread of disease are movement of live fish, discharge of untreated blood and contact with infected vehicles and equipment.

ISA is exotic to the European Union. It is a notifiable disease under UK legislation and a List I disease under European Directive 91/67/EEC. Under EU legislation action must be taken to contain any outbreak, to eradicate sources of infection and to protect other fish farms by:

Compulsory slaughter and disinfection of infected farms

Strict movement controls on suspect farms

Placing farms in the vicinity of an outbreak under surveillance

EPIDEMIOLOGY

In the autumn of 1984, a new disease was observed in Atlantic salmon being farmed along the southwest coast of Norway. The disease, which was named infectious salmon anemia, spread slowly, but caused the death of up to 80% of salmon stock in some hatcheries. By June 1988 it had become sufficiently widespread and serious to require the Norwegian Ministry of Agriculture, Fisheries and Food to declare it a notifiable disease.

In the summer of 1996, a new disease appeared in Atlantic salmon being farmed in New Brunswick, Canada. The death rate of the fish on affected farms was very high and, following extensive scientific examination of the victims, the disease was named "hemorrhagic kidney syndrome." Although the source and distribution of this disease was not known, the results of studies by Norwegian and Canadian scientists showed conclusively that the same virus was responsible for both infectious salmon anemia and hemorrhagic kidney syndrome.

In May 1998, a salmon farm at Loch Nevis on the west coast of Scotland reported its suspicions of an outbreak of infectious salmon anemia. The suspicions were confirmed, and by the end of the year, the disease had spread to an additional fifteen farms not only on the Scottish mainland but also on Skye and Shetland.

In 2008, there was an outbreak of ISA in Shetland. ISA was detected in fish from just one site. There is no evidence the disease has spread beyond this site, but two nearby SSF cages are under suspicion of carrying the disease and are also now clear of fish.

In 2011, two wild Pacific salmon taken from the central coast of British Columbia were suspected to have ISA after preliminary tests showed possible evidence of the virus. However, extensive testing by the Canadian Food Inspection Agency to try to amplify and culture the virus were unsuccessful, prompting the agency to conclude that the ISA virus was not present. In February 2012, a confirmed outbreak of ISA in Nova Scotia resulted in the destruction of up to 750,000 salmon.

In July 2012, a confirmed outbreak of infectious salmon anemia in Newfoundland and Labrador, Canada, prompted the destruction of 450,000 farmed salmon by the Canadian Food Inspection Agency and an outbreak was also confirmed at another site in late 2012. In mid-2012, another outbreak was identified in Nova Scotia, with the 240,000 fish being allowed by CFIA to mature to market size before being harvested in early 2013 by the operator and processed for the consumer market. After being held by the CFIA, the fish was declared fit for human consumption despite the presence of the virus as the disease "poses no risk to human health". In January 2016, it was announced that the virus had been discovered in farmed and wild salmon British Columbia for the first time.

In Chile, ISA was first isolated from a salmon farm in the 1990s and described for the first time in 2001, although the initial presence never resulted in widespread problems. However, since June 2007, the national industry has been dealing with a serious ISA outbreak which has not yet been completely brought under control and has been responsible for an important decline in the industry, closure of many farms and high unemployment. The virus was detected in an Atlantic salmon farm in Chiloé Archipelago in Los Lagos Region and spread to the fjords and channels of Aysén Region to the south almost immediately.

Treatment

There are no treatments for ISA and no licensed vaccines in the EU. Vaccine trials in Canada have yielded equivocal results.

Infectious salmon anemia virus susceptibility and health status of wild versus farmed Atlantic salmon: A comparative study

There are concerns about the potential interaction between farmed and wild Atlantic salmon in areas where they coexist. The Committee on the Status of Endangered Wildlife in Canada (COSEWIC) has designated wild Atlantic salmon populations in Atlantic Canada as threatened or endangered. While the health status and disease resistance of farmed Atlantic salmon is well documented, the information for wild Atlantic salmon is less abundant. For example, Infectious Salmon Anemia Virus (ISAV) remains a recurrent problem for cultured salmon in Atlantic Canada, with outbreaks detected in Nova Scotia and Newfoundland since 2012.

There are knowledge gaps regarding the prevalence of this virus in wild populations as well as the potential transmission between wild and cultured stocks. As such, the aim of this project is to use in vivo disease challenges and next generation sequencing technologies to compare the susceptibility of wild Atlantic salmon stocks (St. John River, Inner Bay of Fundy, Miramichi, Margaree river stocks) and cultured stocks (St. John River origin) to diseases such as ISAV and others. This study will examine genetic differences and measure immune responses, which could potentially explain any observed differences in susceptibility of wild and cultured salmon to ISAV. It will also provide a general assessment of the health status of wild stocks. The evolution rate of ISAV will be examined by looking at full ISAV sequences in tissue of farmed and wild Atlantic salmon sampled throughout the course of the in vivo challenges. This project will address the knowledge gap regarding the susceptibility of wild Atlantic salmon compared to farmed Atlantic salmon to ISAV.

- Aquaculture Collaborative Research and Development Program (ACRDP) Fact Sheet
- Use of Genomics to Control Infectious Salmon Anemia Virus (ISAv)

Infectious Salmon Anemia virus (ISAv) poses a serious threat to fish farmers in Canada and globally. It can cause significant economic losses to Atlantic Salmon farming operations, and is a potential risk to wild Atlantic Salmon populations. Although improved farming and management practices have helped reduce the risk and impact of the disease, further outbreaks associated with new virulent strains remains a threat to the industry. A greater understanding of fish-virus interactions at the molecular level is of critical importance in the development of effective molecular tools and vaccines. A common approach for studying post-pathogen interactions at the molecular level is the use of functional genomic tools, such as DNA microarrays, which measure gene expression in specific tissues at specific times during the infection.

With the use of this method, it has been possible to identify several genes up-regulated or down-regulated in salmon kidney tissue during the course of an ISAv infection and has provided a better understanding of how the fish fight the disease. Gene expression was measured at five different time points during an experimental infection: 6 hours, 24 h, 3 days, 7 d and 16 d after injection of the virus. Several genes involved in immune defense mechanisms were upregulated as the infection progressed. During the last stage of infection, several genes involved in oxygen transport were down-regulated. Such gene expression closely corresponds to anemia (low red blood cell count) observed in Atlantic Salmon infected by virulent strains of ISAv just prior to death. Such studies provide a deeper understanding of the defence mechanisms of fish against disease and provide insight into the approaches that could be taken to help the industry and prevent infections.

Infectious Salmon Anemia virus (ISAv) poses a serious threat to stock health of the fish-farming industry. The virus can cause systemic disease that can lead to mass mortality in Atlantic Salmon (Salmo salar). Infected fish suffer from hemorrhaging, pale gills, congestion of internal organs and severe anemia, often leading to death. Other members of the salmonid family, including Brown Trout (Salmo trutta), Rainbow Trout (Oncorhynchus mykiss) and Coho Salmon (Oncorhynchus kisutch), can also be infected by ISAv. However, the majority of these infections are asymptomatic (a fish may carry the disease but experience no symptoms). Although ISAv has similarities with the influenza virus, it is not zoonotic (humans cannot be infected).

The disease was first reported in Norway in 1984 and has since been detected in Canada's East coast, the United States, Scotland, the Faroe Islands and Chile. The virus was identified in 1996 in New Brunswick, where it devastated the aquaculture industry. A number of biosecurity measures were introduced to limit the spread of the infection, such as aquaculture bay management units, but the risk of new outbreaks remains. Various strains of ISAv have been identified and characterized over the years and show a range of virulence from low to high, depending on the strain.

Although the evolution and pathogenicity of the various ISAv strains have been extensively documented, few studies have been conducted on the molecular interactions that take place between the virus and the host during an infection. When salmonids are infected with a virus, several defense mechanisms are activated, including an immune response and production of antibodies which will immunize the fish against subsequent viral attacks. However, with ISAv this response is often not sufficient to fight the virus, resulting in death of the fish. Understanding how salmon respond to the ISAv during the course of an infection is vital to the development of new antiviral treatments and new vaccines. Given the major economic importance of salmon to the aquaculture industry and the potential disaster that could result from an ISAv infection outbreak, several teams of researchers are trying to develop a vaccine against the virus. When effective, vaccines stimulate the production of specific antibodies and protect the fish against future, related viral attacks. Effective vaccines are available for a number of fish diseases; Vibrio, furunculosis, etc., but to date, salmon have not responded well to the ISAv vaccines in development. The testing of vaccines is a long and costly process and therefore it would be extremely useful if a gene expression profile indicating the onset of the immune response in fish could be identified. This profile would be developed after vaccination and a short incubation period. It is very likely that effective vaccines would produce a genetic response in fish that is very similar to the response shown following an actual ISAv infection.

In recent years, a series of genome projects has led to the sequencing and construction of complementary deoxyribonucleic acid (cDNA) libraries for fish species, including Atlantic Salmon and Rainbow Trout. These cDNA libraries have made possible the development of DNA chips, which have been useful in many functional genomic studies. DNA chips consist of glass slides onto which DNA spots are printed. Each spot represents a unique genetic marker, usually a gene expressed in the host tissue.

The purpose of the present study is to identify genetic markers (a gene or DNA sequence of known location on a chromosome) indicative of an ISAv infection, as it progresses in time. It is assumed that many of these markers would also be present following vaccination with an effective vaccine formulation. An initial test of vaccine formulations could be performed to select the most promising for more detailed testing. The Consortium for Genomics Research on All Salmonids Project (cGrasp) developed a salmonid DNA microarray (or DNA chip) that holds 32,000 DNA spots. These DNA chips were usedto identify genes that were differentially expressed in the kidneys of Atlantic Salmon during the course of an infection with a virulent strain of ISAv.

METHODS

Controlled exposure study

Prior to commencing a challenge, Atlantic Salmon smolts were housed for three weeks in sea water maintained at 13°C in the quarantine laboratory at the St. Andrews Biological Station, New Brunswick. The salmon were injected with a virulent isolate of the HPR4 strain of ISAv and held in duplicate tanks. The controls were injected with culture medium only. The fish were sampled at times 0 hours, 6 hours, 24 h, 3 days, 7 d and 16 d after injection and examined for signs of infectious anemia. Kidney sections were removed from each fish and preserved until analysis.

Sample preparation

Total ribonucleic acid (RNA) was extracted from each kidney sample, purified, quantified and tested for quality control. Reference RNA was prepared from a separate group of healthy fish. The reference RNA represents the normal state of non-infected salmon. Each salmon sampled from the experimental tanks, injected with either ISAv or culture medium (control), was analyzed in comparison to the reference fish. For each sample of infected fish collected during the experiment, RNA was transcribed in complementary DNA (cDNA) and labeled (stained) in vitro with a fluorescent red dye. An RNA reference sample was similarly labeled, but with a green dye. Labeled cDNA from an experimental fish (control or infected) was mixed with cDNA from the reference fish. The mixture was then applied to the cDNA slide (microarray). Each of the 32,000 spots (or groups of identical DNA strands) on the slide represents a short gene sequence. When the slides are exposed to labeled cDNA from the infected and control fish, it hybridizes with the corresponding DNA spotted on the slides.

Scanning of DNA microarrays and image analysis

The DNA microarrays were laser-scanned and digitized for purposes of image analysis. The fluorescence intensity of each spot (gene) on the slide was measured. Fluorescence reflects the contribution of both fish (the cDNA of the reference fish, in green, and the cDNA of the experimental fish, in red). Since each spot on the slide represents a gene, the spots with a green intensity reflect a predominant contribution from the reference fish, or a reduced expression of that specific gene in the experimental fish. Spots with a red intensity reflect a marked predominance of cDNA from the experimental fish or an increased expression of that gene in the experimental fish. Fluorescence intensity indicates the amplitude of that differential expression. Genes that are not differentially expressed in the experimental fish will appear in yellow.

Statistical analysis

Differentially expressed genes were identified for each sampling time point (6 hours, 24 h, 3 days, 7 d, 16 d) by comparing the ratios of ISAv-injected fish to fish injected with culture medium (controls). Significant genes identified for each sampling time point were then categorized by their function.

Validation by qRT-PCR

Quantitative Real-time reverse transcriptase polymerase chain reaction (qRT-PCR) was performed on each fish to verify the infectivity status. The same technique was used to confirm the differential expression of a selection of genes identified by the microarray analysis.

RESULTS

The first mortalities of Atlantic Salmon infected by the ISA virus were observed on day 12 after the injection. On day 16, the end of the experiment, most of the remaining fish were close to death. The controls were healthy and tested negative for ISAv. The virus was detected in the kidney of ISAv-injected fish after 6 hours (3 out of 8 kidneys sampled). After 3 days, all injected fish tested positive and the viral load remained high for the rest of the experiment.

Several hundred genes were differentially expressed during the course of the infection, some observed after 6 hours and 24 h, but most observed at subsequent time points.

Of those genes, a clear over expression of components associated with the development of the innate immune response and acquired immunity were observed. After 16 days, genes associated with an immune response (in this case antigen presentation to cytotoxic T lymphocytes - TCD8), were activated; several genes involved in oxygen transport were underexpressed.

As expected, overexpression of genes involved in the presentation of the virus to immune system cells was observed, but the activation of mechanisms leading to antibody production could not be detected. It is important to note that antibodies provide a long-term defence against subsequent infections. After 16 days, when most fish were dying, the viral infection had overcome the host's defence. Anemia (or blood deficiency) is a characteristic symptom of ISAinfection and the ISA virus is known to adhere to the red blood cells of Atlantic Salmon. The gene response observed at that stage was closely correlated with the symptoms observed in the fish (i.e., underexpression of several elements associated with oxygen transport and red blood cell production).

With the use of DNA microarrays, it was possible to detect the immune defence mechanisms of Atlantic Salmon to ISAv and to describe the sequence of physiological events leading to death. The genetic markers identified can be used to study new vaccines and to evaluate the effect of new isolates of ISAv. DNA microarrays are powerful tools for large-scale measurement of gene expression. The qRT-PCR method enables absolute quantification at a smaller scale and is a method of choice for validating the results obtained using microarrays. Once developed and validated using the qRT-PCR method, the results can be used in other similar experiments.

This experiment revealed that naïve salmon (salmon that have not been used in previous experiments) responded strongly to ISAv infection. It would be interesting to expand the research to include how other variables would be likely to affect the immune response in salmon (e.g., resistance of certain families of fish, the level of virulence of the isolate injected, environmental conditions that affect the disease such as hypoxia and temperature effects). Such studies provide a deeper understanding of the defence mechanisms of fish against disease and provide insights into the approaches that could be taken to help the industry and prevent infections.

Area management strategies for disease control in aquaculture: Perspectives on infectious salmon anemia (ISA) spread from two continents

The coordination of production and biosecurity practices across neighboring farms in a geographically, or hydrographically, defined region is an important disease prevention and control strategy in aquaculture. Here termed area management, this fairly rigid organizational structure requires buy-in and coordination at many levels, both within and between companies, as well as among local, national and, at times, international governments. Area management in aquaculture perhaps stood its most prominent test in its widespread adoption in response to outbreaks of infectious salmon anemia (ISA) internationally. Successes in ISA virus (ISAV) control have since been attributed to the area separation achieved through synchronized stocking, fallowing, movement restrictions, and fomite or pest control.

Area management, however, is costly; often demanding extra biosecurity, lengthy or inconveniently timed fallows, and localized equipment, personnel, and services. It is also inefficient, generating redundancies in workforce and infrastructure, and periodic lapses or surplus of product that may not sync with market demand. Yet high in both cost and potential impact, few epidemiologic studies have examined the utility of area management methods directly. ISA upended Atlantic salmon production in the east coast of the United States in 2001/2002. Adoption of area management practices in Maine, and harmonization of the same with New Brunswick across the international border, is generally credited with the fairly rapid resolution of the outbreak.

The importance of waterborne spread of virus in and between salmon farming areas in Maine and New Brunswick was demonstrated in retrospective analysis, and similar analyses suggested also the importance of fallows. However, data on additional area management tenets lacked volume, variability or controls for empirical analysis. More recently, Chile's National Fisheries and Aquaculture Service implemented area management practices in response to their 2007 emergence of ISAV. Longitudinal data collected throughout this outbreak allowed the evaluation of the impact of additional area management tenets on virus control. In Chile, retrospective analysis identified hydrographic linkages, shared ports, and fish transfers from areas with recent occurrence of ISAV as the strongest predictors of virus spread between areas, providing support for their strong relevance to area management success.

We compare patterns across continents, and discuss observations and questions that could impact future disease control strategies. For example, in Chile, hydrographic linkages were most predictive of disease spread prior to national adoption of enhanced biosecurity and fallowing regulations, suggesting that viral load can impact spread dynamics. As another example, results varied by ISAV type (generally categorized HPR0 for non-virulent genotypes, and HPRv otherwise). In both countries, HPR0 arose late in the study period, so few HPRv events were available by which to explore the hypothesis of HPR0 as progenitor of outbreaks.

However, spatiotemporal patterns in the occurrence of both types suggest a parallel or possibly reversed means of spread. In Chile, the occurrence of HPR0 could be predicted in part by the distribution of HPRv events almost 3 years prior. Finally, we highlight the importance of ongoing efforts to improve data precision, breadth and consistency in field settings; and the importance for continued HPR0 study, both to specify its diagnostic test accuracy and define its role in HPRv outbreaks. These and related efforts will serve to better inform efficient and effective area management strategies.

Infectious salmon anemia a grave threat to fish farms

Infectious salmon anaemia (ISA) is a viral disease that affects Atlantic salmon but poses no threat to humans. First identified in Norway in 1985, it was detected in Canada in 1997 but was believed to be "exotic" in the EU until 1998 when it was discovered in a salmon farm at Loch Nevis.

A report published in Britain last year was unable to identify the origin of the outbreak, but narrowed it down to two possible causes - accidental introduction on a boat from infected areas in Norway, or introduction through fish in a wild reservoir.

Initially there was widespread slaughtering of fish in Scotland, particularly in the Shetlands, as it was believed that the disease could only be contained by removing and destroying all the fish from the infected site, disinfecting equipment and fallowing the farm for at least six months.

At the time, it posed a serious threat to an industry employing some 6,000 people with a turnover of £300 million in 1998. The condition spread to over 30 farms in Scotland, but many of these sites are now back in production after fallowing for the required period.

In 1991 ISA was classified as a List 1 disease, with control measures requiring the immediate withdrawal of fish, but the European Commission took a more pragmatic approach in 1998. This allowed for efficient control of the disease while also trying to safeguard the interests of fish farmers. Where a remote farm had low mortalities and few clinical signs of the condition, a cage-by-cage clearance programme was permitted. However, infected farms close to other fish farms and experiencing high mortalities were required to clear as soon as possible.

Dr Ken Whelan of the Marine Institute said: "Obviously we have be very careful in relation to wild fisheries, particularly when there has been an escape, but so far we have no clinical outbreak here and the Marine Institute, North-Western and Western Regional Fisheries Board are working together to remove any of the escaped rainbow trout from the Clew Bay farm that are detected."

The Global Spread of Infectious Salmon Anemia

As a second wave of Infectious Salmon Anaemia washes up on Scottish shores, the industry holds its breath. Only time will tell what gets upturned in its wake, but on a global scale the severity of the problem is already showing its true face, writes Adam Anson, reporting for TheFishSite.

Almost immediately after the virus was reported in the Shetlands, fish from infected farms were removed and inspectors were making their way around the 42 fish farming sites within the newly established Control and Surveillance Zones. The efficiency of the reaction belies the severity of the problem, and may come to save this industry from the graver impacts that have come to dominate salmon fisheries elsewhere in the world.

It will take approximately six weeks to culture the samples and establish definitively the presence of the virus, but the first indications of the results should be known much sooner. The Scottish Industry has also initiated an investigation into the source of the outbreak, saying it will be "the subject of a scientific study to determine the source of this new infection, the distribution of the disease in the environment and the risk of further spread."

There has to date been one previous outbreak of ISA in Scotland, which occurred over the years 1998-99. According to figures from the Fisheries Research Service (FRS), the estimated cost to the industry was in the region of £30 million. In this initial incident prompt action succeeded in eradicating the disease, but other countries have been less fortunate.

Since it was first detected back in Norway in 1980's, ISA has posed a number of resilient problems for salmon farmers. To this day, it still hampers the Norwegian industry.

In the same year that Scotland managed to eradicate ISA for the first time, the average annual cost of the epidemic in Norway was US$11 million, while in Canada the virus was estimated to cost an average of US$14 million per year. The Faroe Islands paid the price of their entire industry when the virus hit in 2000, while the Chilean industry currently toils from a particularly severe series of ongoing outbreaks. Chilean salmon exports are currently one of the country's most dynamic businesses, boasting over 800 salmon farms, with an annual turnover of more than US$2 billion. However, since the 2007 more than 100 Chilean fish farms have been affected. According to reports, approximately two per cent of farmed Chilean salmon die as a direct result of the virus.

The industry has since deteriorated drastically and consumer confidence is falling. SalmonChile reported that production could drop by as much as 20 per cent in 2009 – possibly to as low as 275,000 tonnes (13/10/08), a far cry from its heights in 2007 - when it was estimated to have produced just shy of 400,000 tonnes of salmon and reported export earnings of US$2.247 billion.

The Chilean salmon industry portrays the depth of this problem and there appears to be no solution in sight. Far from coming under control the virus continuously alludes preventative measures, passing through the safety net of the most stringent control measures in the world. According to figures from the World Animal Health Information Database, over the course of 2008 the ISA virus claimed an outbreak in Canada, 20 outbreaks in Chile and 14 outbreaks across Norway. Now, with the inclusion of Scotland, all the big salmon farming countries have recently suffered.

The One that Got Away

Based on biochemical, physical, chemical, and structural characterization, ISA virus is very similar to members of the "flu" family, says the United States Department of Agriculture (USDA). Although there are no human food safety concerns, the mortality rates of salmon infected are highly variable, ranging from two to 50 per cent over one production cycle, but potentially affecting an entire farm in a matter of months.

Signs of infection in salmon include: lethargy; loss of appetite; gasping at water surface; pale gills; dark liver; accumulation of fluid in the body cavity; hemorrhage in internal organs and high of levels of mortality, but the global impact of the virus shows not only how devastating it can be, but also how difficult the problem can be to eradicate once it has firmly set in.

The World Organisation for Animal Health (OIE) says that vaccination against ISA was carried out in North America over five years, but the available vaccines did not offer complete protection. Similarly, the use of antibiotics is another ineffective solution and is one of many other factors that could lead the industry into deeper, longer lasting troubles.

There is evidence that poor sanitation, unnatural growth conditions, and environmental pollution are at the root of this problem. ISA is thought to affect fish in stressful conditions in highly increased rates. It has also been suggested that over crowded pens placed in close contact with one another are adding to the weight of this problem, whilst making it easier for the virus to harbour in large numbers and pass from one farm to the next.

The most effective way to combat the disease may well be with natural ecologically sound measures, quick response to outbreaks, removal of fish from site, sound disposal, clean farming conditions and well managed operations, but the cost to create these conditions will arguably be high and restructuring well rooted industries will not be easy to manage.

Whether the cost of implementing such procedures balance out in comparison with the current price of salmon on the market is yet to be realised.

FISHING FOR THE SOURCE OF THE PROBLEM

Despite of previous investigations it has been difficult to pinpoint the exact factors pertaining to these outbreaks and even to source where the virus originated.

According to the FRS of the Scottish Government, the ISA virus is spread through water. Once infection is established in a farm, other farms in the area are at risk. The risk of transmission is associated with infected biological materials such as animal wastes or discharges from normal culture operations and "untreated blood released into the sea", but there has also been evidence of "vertical transmission" via eggs or milt, says the FRS. It has also come to light that "other species of fish may be carriers" also. However, although other fish can harbour the disease, it only really affects the salmon industry. The FRS say that the extensive testing of wild fish at the time of the outbreak in 1998/99 did find some evidence evidence of ISA virus in salmonid species, but prevalence of infection was low and there was "no evidence that this caused disease and no observations of clinical signs". There is no evidence from Norway, where ISA has been present since 1984, that ISA causes disease or harmful effects in wild Atlantic salmon at all.

According to a Geologic Survey conducted by the National Fish Health Research Laboratory, sea lice may play an important role as vectors that can enhance contagion during epidemics and evidence has pointed to certain species of fish and shellfish that can either accumulate the pathogen or transmit the disease. This research indicates that wild fish may become carriers and serve as potential reservoirs of infection while the first Scotland outbreak proved inconclusive, it raised many questions about how ISA managed to penetrate the industry. Consequently it led to many perceived solutions that have since been implemented. The Scottish government says that importation of live salmonids is normally prohibited except from areas of equivalent health status and imports of live fish into Scotland have not been allowed from Norway, Chile or Canada as no salmon populations there satisfy the fish health requirements for import into the UK. Yet, with this second bout of infection, new questions are bound to be raised.

One report to come from the Fisheries Research Services Marine Laboratory in Aberdeen identified the critical role that the long-distance transport of pathogens plays in the emergence of novel diseases. Shipping is a major contributor to such transport, and the role of ships in spreading disease has been recognized for centuries.

"We present evidence of invasive spread of infectious salmon anemia in the salmon farms of Scotland and demonstrate a link between vessel visits and farm contamination", claimed the report..

Whatever the outcome, the source of these outbreaks must be traced and understood if the industry is ever to close it and secure their operations from this disease. Perhaps only then will the long apprehensive wait of salmon farmers in the Shetlands and the world-over ever come to an end.

Long-distance transport of pathogens plays a critical role in the emergence of novel diseases. Shipping is a major contributor to such transport, and the role of ships in spreading disease has been recognized for centuries. However, statistical confirmation of pathogen spread by shipping is usually impractical. We present evidence of invasive spread of infectious salmon anemia in the salmon farms of Scotland and demonstrate a link between vessel visits and farm contamination. The link is associated with vessels moving fish between sites and transporting harvest, but not with vessels delivering food or involved in other work.

Human Development and Infectious Disease in Coastal Ecosystems

Anthropogenic activity increases the incidence of infectious diseases, which in turn influence the populations and production of marine organisms, from free-living bacteria to mammals. The aquaculture industry has been strongly affected by diseases emerging from anthropogenic activities, and itself has played a critical role in their spread.

Infectious salmon anemia is an emerging disease causing severe damage to the salmon-farming industry in an increasing number of countries. The disease, first reported in Norway in 1984, has since been reported in Atlantic Canada (1996), Scotland (1998), the Faroe Islands and possibly Chile (1999) , and most recently Maine, USA (late 2000), over the last few months (2001), infectious salmon anemia has spread rapidly in Maine. In 1999, the annual cost of infectious salmon anemia was reported to be US$11M in Norway and US$14M in Canada, while in Scotland, the total cost of the epidemic of 1998-99 was US$32M. It is too early to say what the cost of the disease will be in Chile and the United States, but both countries have large salmon-farming industries. In almost all cases, infectious salmon anemia has mainly affected Atlantic salmon, Salmo salar, but in Chile, deaths have been reported among Coho salmon, Oncorhynchus kisutch. All deaths have been among farmed salmon.

Infectious salmon anemia is an emerging disease caused by novel virulent strains of a virus that has adapted to intensive aquacultural practices and has exploited the associated traffic to spread both locally and internationally. Genetic and phenotypic differences suggest that this adaptation occurred independently in Europe and the Americas. The virus strains then aggressively expanded their geographic ranges.

Infectious Salmon Anemia Virus (ISAV) in Scotland

In May 1998, the previously exotic viral disease infectious salmon anemia was detected at a salmon farm in Loch Nevis on the Scottish west coast.ISAV is an othomyxovirus of a new genus (Aquaorthomyxovirus) that is closely related to the influenza viruses ISAV subsequently spread to salmon farms throughout Scotland. Molecular epidemiologic studies indicate that a single strain was responsible for the initial epizootic, although a second strain unrelated to the outbreak was later identified.. Infectious salmon anemia has been reportable in the United Kingdom since 1990, so its spread has been well surveyed and documented. The pattern of spread was discontinuous, with farms in Shetland becoming infected during summer 1998 but ISAV detected in Orkney in late 1999. Many intervening areas did not become infected, and in many cases, farms close to infected sites did not become infected. The broad area of infection with multiple isolated foci is not consistent with diffusive spread of disease by fish-to-fish contact or through vectors such as parasites or seabirds, this is not to say that this form of spread may not have occurred occasionally or locally in the vicinity of outbreak foci.

Invading new areas is critical for the survival of exotic species including pathogens. Shipping has been identified as a major factor in movement of exotic species to coastal regions. The role played by ships in the introduction and spread of Black Death (a virulent form of plague) in 14th-century Europe has been extensively chronicled .Recently, huge numbers of bacteria (8.3 x 10 l), including Vibrio cholerae, the agent of cholera, and viruses (7.4 x 10 l) have been detected in ballast water of ships entering U.S. waters . Given the increasing volume of shipping introduction of pathogens to coastal ecosystems is likely to increase. The rapidly growing aquaculture industry, with its high densities of potential host monocultures, is based in such coastal ecosystems

Numerical analysis of the role of shipping in spreading pathogens is usually not possible because of heavy ship traffic and the multitude of pathogen sources. In this article, we examine the role of shipping in the invasive spread of infectious salmon anemia among Scottish salmon farms.

Because of the broad pattern of spread, the movement of live fish and contaminated equipment between fish farms was suspected to play a role in the spread of this disease within Scotland. In aquaculture, salmon are moved extensively. They are reared at freshwater hatcheries, transferred to (and sometimes between) marine production sites, and finally transported to central harvesting and processing stations. In particular, well boats, which are used for the transport of fish, supplies, and equipment between farms and as work platforms, were suspected in the spread of ISAV.

Well boats and ferries have been suggested as possible routes of introduction of otherwise unexplained outbreaks of infectious salmon anemia in salmon farms in Norway, where the disease has been established since 1984 and has infected 101 fish farms in 1990 alone.

We conducted a quantitative analysis of the spread of ISAV in Scotland and its relationship to the movements of four well boats that serviced the farmed-salmon harvesting center of Loch Creran in the year preceding the epidemic (May 1997 to May 1998). These well boats made 1,558 visits to fish farms along >850 km of coastline from Loch Fyne to Shetland. Farms not visited by these well boats are not included in this analysis (including in Orkney and the Western Isles, where ISAV was later detected).

These farms were serviced by other well boats, for which movement data were not available. We grouped the visited farms into 26 areas since individual fish farms are not always identifiable from the boats' logs. Of these areas, 6 were infected, 7 were suspected of being infected, and 13 escaped infection. Infectious salmon anemia was confirmed if clinical disease was present and ISAV was identified as the causal agent by both pathologic lesions and the presence of virus. In areas where infection was suspected, ISAV was identified by polymerase chain reaction (PCR) analysis ,or similar immunofluorescence, methods or by clinical signs without confirmation of infection. A histogram of the number of areas included in these three categories clearly showed a relationship between the log number of boat visits and the risk for infection.

In two areas, infection developed after only a few visits. Loch Broom was suspected of being infected after a single visit by one of the boats. This infection occurred in 1999 and was therefore not part of the initial invasion. Infection also occurred on the Isle of Skye, where only two well-boat visits were recorded; however, in this case, fish stocks were moved directly from the infected site at Loch Nevis before infection was detected. Similarly, infectious salmon anemia was introduced to Shetland. when a partial load of smolts was brought to the Skye site, contaminating the load intended for shipment to Shetland. Since were were direct introductions of infected fish to these sites, we excluded them from most analyses.

The harvest-processing station at Loch Creran used by the four vessels is also excluded from further analysis. Visits to this processing station, where fish were unloaded and held temporarily in net pens before slaughter, were qualitatively different from harvesting visits to other sites, where fish were grown. The fish at Creran were from multiple sources, while fish at other sites were from one or a few sources. Because all harvest trips ended at the processing plant, its inclusion would lead to double counting of harvest trips. The exclusion of Creran is particularly important for analyzing the efficiency of infection transfer during different types of visits.

The multivariate analysis shows that infection status is not related to distance from Loch Creran or movement of well boats for general purposes. Infection is transferred only by the shipment of live fish or visits to the harvesting site. If the sites infected by the movement of live fish from Loch Nevis and the uncertain site of Loch Broom are excluded, fish movements are also not related to infection status. Infected sites whose cause of infection is not explained by direct movement of infected fish are thus very strongly related to the number of harvest visits ($p = 0.0009$).

Harvesting involves transfer of live fish to Loch Creran, which is probably how this harvest-processing station became infected. At the time, the processing plant adjacent to the harvesting site was discharging effluent that had not been disinfected. Movement of the well boats from Loch Creran may then have rebroadcast ISAV.

Analysis of regional infection patterns within 50 km of Loch Creran shows a pattern of infection similar to the national pattern in that only movement of harvest vessels correlates with area infection status and r is similar at 0.62. However, the relationship is far weaker, only 12 points ($p = 0.049$), because fewer data are available. The regional pattern of infection excludes, among other areas, Skye, Shetland, and Loch Broom. At this regional scale, the processes and sensitivities of transmission do not appear to differ much from those at the national level.

At the smaller scale of spread of within a few kilometers, which the coarsely resolved available data cannot resolve, other processes may have become important, including physical transport of virus by currents (associated with waste products,escaped or wild fish ,or vectors (such as sea lice) , moving between neighboring farms. Infected wild salmonids, which have been found throughout Scotland could act as an ISAV reservoir.

The salmon farming industry and regulatory authorities responded to the epizootic by introducing strict controls on hygiene and on movements of well boats and live and dead fish and by making the slaughter of infected stocks compulsory. These measures resulted in the closure of the Loch Creran site (among others) and control of the major route of infection. In 1999, sites not visited by these well boats (and also Loch Broom, which was reportedly only visited once) became ISAV suspect. It is possible that once ISAV had become widely distributed, vessels other than those based at Loch Creran transported ISAV to new locations. Other transmission processes (e.g., sharing divers or equipment) could also have played a role in disseminating the virus.

The evidence presented here supports a very strong quantitative link between the number of visits by well boats and the probability of ISAV detection in an area. The pattern of spread does not support a natural diffusive expansion of the ISAV epizootic, but the identical genetic nature of the ISAV at different sites indicates a direct link between incidents. Management activities could have resulted in interregional contact that coincided with the number of visits by well boats, but the strength of the relationship between well-boat visits and infection implies that the well boats played a predominant role in transmission. This role is emphasized by the relationships between the specific type of visit and infection status. Simple exchange of equipment does not appear to increase risk for ISAV infection in Norway and neither do general well-boat visits in Scotland. We therefore conclude that it is the movement of well boats, through shipment of live fish and visits for harvesting, that spread ISAV at the regional and larger (ranging from 10 to several hundred kilometers) scale in Scotland during 1998-99.

No new incidents of clinical infectious salmon anemia occurred in salmon farms in Scotland in 2000, and infection did not develop in new areas, although more farms reported isolated ISAV-positive reverse transcription PCRs. ISAV also emerged in or was spread to the North Atlantic Faroe Islands. ISAV (but not infectious salmon anemia) was detected by PCR methods in wild salmonids throughout Scotland in 1999 , but although present, was much less prevalent in 2000 . The rapid response of the industry controlled the disease in farmed salmon and may have prevented endemic infectious salmon anemia in Scotland as it has in Norway.

After the introduction of stricter regulations in Norway in 1990 and 1991 to control hygiene in vessels and slaughterhouses, infectious salmon anemia outbreaks were reduced by nearly two orders of magnitude. This reduction shows that similar transport of poorly sanitized material by ships was also an important mechanism behind the infection's spread in Norway. However, infectious salmon anemia is established in Norway, and since 1991, its incidence has gradually increased again.

Our data suggest that well boats have played a major role in the spread of infectious salmon anemia in both Scotland and Norway. The pathogens carried by well boats had two possible sources: the processing plant, when the wells were unloaded and replacement ballast water was contaminated with processing plant effluent, or the adjacent harvesting station, when infected water was taken up. Alternatively, infected fish or fish detritus may have remained in the wells or pumps and pipework after fish were discharged to the harvest station and the vessel immediately left for a harvesting operation at another site. When well boats transfer between sites, they are inspected and disinfected, which should minimize the risk for infected cargo residue transferring ISAV. However, removal of all fish and residue from pumps and pipework is problematic. At the time of the epidemic, effluent from the Loch Creran processing plant was not fully disinfected, so ISAV could have been present in ballast water taken up after disinfection.

When the well boat starts harvesting, this water is discharged so that fish can be loaded. It is difficult to assess the role of the two sources of pathogens in the shipborne transmission of infectious salmon anemia, although since pathogens are abundant in ballast water ,the role shown for the harvesting station in broadcasting ISAV could have wider implications for the transmission of diseases by shipping. Infectious salmon anemia is only one of many emerging diseases present in marine environments, the aquaculture industry can both suffer from and assist in the spread of such diseases. While individual movements of potentially infected fish, particularly live stocks , processed or unprocessed carcasses used for fish food , or fish imported for human consumption , are most likely to spread disease, these movements can be monitored and controlled. Extensive ship traffic and lack of regulation increase the risk of spreading disease to animals raised for aquaculture and to other animals in marine environments.

Diseases potentially spread by shipping include waterborne diseases of humans such as cholera and potential viral zoonoses. Thus, although this article highlights anthropogenic spread of a pathogen economically damaging for aquaculture, it also underscores the potential role of shipping in the global transport of zoonotic pathogens.

INFECTIOUS SALMON ANAEMIA VIRUS IN WILD FISH FROM SCOTLAND

Following the outbreak of infectious salmon anaemia (ISA) at salmon farms in Scotland, UK, a survey was established to determine the extent of infection in wild fish. All fish tested were free from the clinical symptoms of ISA. Isolations of ISAV were made from 5 sea trout within areas where ISA affected salmon farms were located. Evidence for ISAV in other sea trout was provided by ISA RT-PCR diagnostic tests. Results from ISA RT-PCR tests reveal evidence for ISAV being present in salmon parr, adult salmon and juvenile brown trout in rivers distant from salmon farms and indicate that, at the time of the survey (1998-1999), ISAV may have been widely distributed. Nucleotide sequence analysis of segments 2 and 8 showed that for most sequences from wild fish there was 100% homology with ISAV isolated from clinically affected farmed fish although evidence is presented which indicates variability in ISAV sequences from wild fish. Modelling the RT-PCR findings indicates that ISAV among salmonid fish was spatially non-random. Brown trout, sea trout and salmon (adult and parr) show a pattern of occasionally large numbers of positive samples against a background of very low numbers.

Disease in Salmon Aquaculture: Infectious Salmon Anemia Virus

While there are several diseases that impact Atlantic salmon aquaculture, Infectious Salmon Anemia Virus (ISA or ISAv) is the most influential disease. ISA first appeared in Norwegian salmon farms in 1984, and have since impacted farms in the Faroe Islands, Scotland, Chile, the US, and Canada (Asche 406). ISA is a highly contagious disease, previously known as Hemorrhagic Kidney Syndrome, in which affected fish experience anemia, or a lack of red blood cells.

The disease initially has a low mortality rate and can be difficult to detect; however, if the disease is not treated, mortality rates can increase to anywhere between 90 to 100%, potentially wiping out entire farms (CFSPH 1). Currently, there is an incomplete epidemiology, where even the means of transmission is uncertain. The disease is spread through horizontal transmission (from one member of species to another in which there is not a parent-offspring relationship), likely through gill to gill interactions, but possibly through either ingestion or parasites (CFSPH 2).

The disease can also be spread asymptomatically (without showing symptoms) from one Atlantic salmon to another or even through other salmonids such as different species of salmon and trout (CFSPH 2). Because the disease is spread through horizontal transmission, regardless of the actual means of transmission, the transmission rate would increase when organisms are in close proximity to one another, such as within a tank or net pen as seen in aquaculture. Despite the incomplete epidemiology, ISA can be treated using chemical disinfectants or prolonged exposure to extreme pH levels, temperatures, or ultraviolet radiation (CFSPH 5). Furthermore, Aldrin et al. show that areas with high concentrations of fish farms that are affected are likely due to transmission of the disease from neighbors.This demonstrates that ISA is not only difficult to combat in a single farm, but the outbreak affects farms around it as well.

Infectious Salmon Anaemia - Fact Sheet

Infectious salmon anaemia (ISA) is a finfish disease caused by a virus that belongs to a family of viruses called Orthomyxoviridae. ISA is a reportable disease in Canada. This means that anyone who suspects ISA in finfish that they own or work with is required by law to notify the Canadian Food Inspection Agency (CFIA).

What species of finfish can be infected with ISA?

Species known to be susceptible to ISA include:

- Oncorhynchus mykiss (rainbow trout)
- Salmo salar (Atlantic salmon)
- Salmo trutta (brown trout)

Species that may be susceptible include:

- Salvelinus alpinus (Arctic char)
- Clupea harengus (Atlantic herring)

Is ISA a risk to human health?

No. There is no evidence that ISA can be transmitted to humans.

Has ISA been found in Canada?

Yes. ISA was first detected in the Atlantic area in 1996. It has been reported in New Brunswick, Nova Scotia, Prince Edward Island and Newfoundland.

ISA has not been found in the Pacific Ocean watershed or the Pacific Ocean off British Columbia.

What are the signs of ISA?

ISA outbreaks are most common in susceptible farmed finfish reared in seawater. The disease has not been seen at the embryo stage, but can affect the fish at any life stage after hatching. It tends to be a slow moving disease so the death rate may initially be low and increase over time (weeks to months). Depending on the strain of virus, outbreaks of ISA can cause death rates of up to 90 per cent in affected populations of finfish. On average, mortality on any given farm is 30 per cent.

Infected finfish may exhibit any of the following signs:

- loss of appetite;
- abnormal swimming patterns (slow swimming, swimming slowly at the surface);
- fish gasping at the surface;
- grey gills;
- swollen abdomen;
- areas of bleeding that may be present along the belly and sides of the fish;
- dark and swollen kidney, liver and spleen;
- areas of pinpoint bleeding in the fatty tissue surrounding organs;
- areas of bleeding in the pyloric cecae, intestines and liver;
- pale heart; and
- bloody fluid in the abdominal cavity and around the heart.

What is required to confirm a diagnosis of ISA?

When ISA is suspected, the CFIA launches an investigation to collect all relevant information, such as signs of illness in the fish and previous disease outbreaks in the area. At the same time, samples are collected and sent for laboratory testing using validated protocols at one of Canada's three National Aquatic Animal Health Laboratories.

How is ISA treated?

There are no treatment options currently available for ISA. There is a vaccine available in Canada to prevent the disease.

How is ISA spread?

ISA may be spread between finfish by contact with:

- excretions and/or secretions from infected finfish; or
- water contaminated with the virus.

People can spread ISA by moving:

- infected live or dead finfish;
- contaminated equipment; or
- contaminated water.

What measures can be taken to prevent the spread of ISA?

Anyone who frequently handles or works with farmed or wild finfish should:

- be aware of ISA occurrences in the local geographic area;
- be aware of the signs of ISA;
- not use finfish bought in a grocery store as bait for catching finfish or other aquatic animals; and
- dispose of all waste from cleaning and gutting finfish in municipal garbage.

Owners of aquaculture facilities are encouraged to implement biosecurity practices that will minimize the introduction and spread of ISA between production facilities. Finfish handlers must apply and comply with all federal, provincial and/or territorial permits or licences that may be required for the purpose of disease control prior to import into or relocation of finfish within Canada.

How does the CFIA respond if ISA was suspected in an aquaculture facility?

If ISA is suspected, the CFIA places movement controls on the facility to control movement of fish, vessels and/or vehicles, and equipment on and off the site. In these circumstances, finfish, vessels, vehicles and equipment could only be moved into or out of the facility with CFIA approval.

If ISA is confirmed, the CFIA controls its spread by carrying out disease response activities. These activities vary depending on the situation and could include:

- controlling the movements of infected animals;
- overseeing the humane destruction of infected animals;
- verification of cleaning and disinfection activities; and
- requiring the facility to be left empty for a period of time (fallow period) prior to restocking.

What is ISAv, how does it affect fish, and what are the risks?

Question: What is ISAv?

ISAv is an orthomyxovirus (related to influenza viruses) that is thought to have evolved in Atlantic salmon; the first incidence of disease was reported among salmon farms in Norway in 1984. The virus is extremely infectious and can result in high mortalities (in excess of 70%). The main route of infection is "horizontal" (fish-to-fish), though it is also suspected that ISAv can be passed "vertically" (through the eggs) because it was introduced into Chile, where no adult salmon were imported or previously existed. ISAv may also infect herring, cod, Pollock, sablefish, and other marine species. These species, which do not typically show signs of the infection, serve as a "reservoir" that makes eliminating the virus, once present, impossible because they can shed virus that will re-infect salmon.

How is ISAv introduced, and how does it spread?

ISAv can be introduced by the shipment of infected salmon tissue and eggs; water containing the virus could also be infectious. Since it emerged in virulent form from a native strain found in Norway, ISAv has been introduced by the salmon aquaculture industry into Ireland, Scotland, the Faroe Islands and Chile. The detection of the European strain in eastern Canada is likely also due to the movement of infected fish or eggs, although it is possible that is was brought to eastern Canada by migrating Atlantic salmon.

The primary method of ISAv transmission is horizontal (fish-to-fish) because infected fish shed the virus in mucous, urine, feces and ovarian fluid. Sea lice have been implicated as a vector for ISAv, and the wounds they make on a salmon's skin may allow the virus to infect new hosts. It also appears to spread vertically (from an infected female to her eggs), although it is unclear if it is simply found on the outside of eggs or whether it manages to infect the young, unhatched offspring.

Was ISAv introduced to the Pacific Northwest by the salmon aquaculture industry?

At present we do not have enough information to determine if the recent detections of ISAv are from European strains, which would suggest introduction by the salmon aquaculture industry. However, once the recently detected virus is sequenced, we will be able to identify its origin. It may be that a strain of ISAv has been present in the Pacific Northwest for some time, or even be native.

If ISAv is a native virus, why should we be worried about it?

Even if the strain of ISAv found in the Pacific Northwest is native there is still cause for concern. In locations where intensive Atlantic salmon aquaculture is found, highly virulent strains of ISAv have developed (see Can ISAv evolve and what are the implications?). In the Pacific Northwest, ISAv could develop into a virulent form, likely first in Atlantic salmon in aquaculture facilities, and then could spread to native Pacific salmon stocks - potentially from California to the Bering Sea, jeopardizing populations that are already endangered or in decline. The salmon in the net pens would serve as a continuous source of viral particles, creating a long-term exposure to ISAv for native salmonids (salmon and trout) and other marine fishes. Crowded conditions such as those found in aquaculture facilities would allow the virus to mutate and become infectious for new species.

Is there more than one type of ISAv?

There are several "strains" (variants) of ISAv. These have previously fallen into two groups, based on their genetic similarity: one in the North Atlantic (Europe) and the other on the Atlantic coast of North America (although the European strain has also been found in eastern Canada). The virus was also introduced into Chile, where there are no native salmon, on several different occasions by the salmon aquaculture industry. The introduction of a Norwegian strain in 2007 led to disastrous losses in Chilean aquaculture, and killed both net pen Atlantic and coho salmon. Some of those fish had recently been infected by another pathogen (Piscirickettsia, a bacterium), and both may have played a role in the high mortalities observed.

How do the strains of ISAv differ?

The original strain of ISAv is thought to have a full-length "highly polymorphic region" (HPR), which is a gene segment associated with virulence; this strain (HPR0) is not commonly associated with disease. This strain was naturally occurring in the North Atlantic and only resulted in disease after Atlantic salmon net pens were established on the Norwegian coast. Other strains with higher virulence (the ability to spread and cause disease, "pathogenic") have mutations in this region that result in shorter HPR gene segments- these are usually referred to as HPR1, HPR2b, etc. Like many RNA viruses, ISAv mutates very quickly and mutations in other regions may also be important.

Can ISAv evolve and what are the implications?

One of the key features of ISAv is its ability to mutate rapidly and increase its virulence; there is evidence that this has happened independently on several occasions in association with fish farming around the world. We now know that mutation events that lead to a shortening of the highly polymorphic region (HPR) can result in increased pathogenicity. We also have evidence that ISAv can evolve virulence for Pacific salmon species (in the 2007 Chilean outbreak, farmed coho salmon were also infected and killed by ISAv; however, other infections may have contributed to the high mortality). Since all salmonids (salmon and trout) are relatively closely related genetically, it is possible that a chronic exposure to ISAv in intensive culture conditions would eventually lead to the emergence of a more virulent form of the virus that could infect native Pacific salmon species.

Protective oral vaccination against infectious salmon anaemia virus in Salmo salar.

Infectious salmon anemia (ISA) is a systemic disease caused by an orthomyxovirus, which has a significant economic impact on the production of Atlantic salmon (Salmo salar). Currently, there are several commercial ISA vaccines available, however, those products are applied through injection, causing stress in the fish and leaving them susceptible to infectious diseases due to the injection process and associated handling. In this study, we evaluated an oral vaccine against ISA containing a recombinant viral hemagglutinin-esterase and a fusion protein as antigens. Our findings indicated that oral vaccination is able to protect Atlantic salmon against challenge with a high-virulence Chilean isolate. The oral vaccination was also correlated with the induction of IgM-specific antibodies. On the other hand, the vaccine was unable to modulate expression of the antiviral related gene Mx, showing the importance of the humoral response to the disease survival. This study provides new insights into fish protection and immune response induced by an oral vaccine against ISA, but also promises future development of preventive solutions or validation of the current existing therapies.

Transmission

The virus is spread by contact with infected fish or their secretions, or contact with equipment or people who have handled infected fish. The virus can survive in seawater, so a major risk factor for any uninfected farm is its proximity to an already infected farm.

The Lepeophtheirus salmonis sea louse, a small crustacean parasite that attacks the protective mucus, scales and skin of the salmon, can carry the virus passively on its surface and in its digestive tract. Under laboratory conditions Lepeophtheirus salmonis can passively transfer ISA. It is not known whether the ISA virus can reproduce itself in the sea louse, although this is a possibility as viruses can use secondary vectors for transmission, for example Arboviruses such as dengue fever, West Nile virus, or African swine fever virus.

14

BLUETONGUE DISEASE

Bluetongue disease is a noncontagious, insect-borne, viral disease of ruminants, mainly sheep and less frequently cattle, yaks, goats, buffalo, deer, dromedaries, and antelope. It is caused by Bluetongue virus (BTV). The virus is transmitted by the midges Culicoides imicola, Culicoides variipennis, and other culicoids.

Bluetongue is a viral disease affecting sheep, cattle, deer, goats and camelids (camels, llamas, alpacas, guanaco and vicuña). Although sheep are most severely affected, cattle are the main mammalian reservoir of the virus and are very important in the epidemiology of the disease. Bluetongue is a notifiable disease in the UK and suspected cases must be reported immediately to the Divisional Veterinary Manager at the local Animal Health Office. There are many different strains (serotypes) of Bluetongue with each given a number (currently up to 27). Animals that recover from infection with one serotype will be immune to that strain but not to others. The virus is transmitted by biting midges of the genus Culicoides and not normally from direct contact with infected animals. Peak midge populations occur during the late summer and autumn in Europe and therefore this is the time when Bluetongue is most commonly seen. The midges can be carried very large distances on the wind (over 200km) and this has been the primary way Bluetongue serotypes are introduced into new areas.

SIGNS

In sheep, BTV causes an acute disease with high morbidity and mortality. BTV also infects goats, cattle, and other domestic animals, as well as wild ruminants (for example, blesbuck, white-tailed deer, elk, and pronghorn antelope).

Major signs are high fever, excessive salivation, swelling of the face and tongue, and cyanosis of the tongue. Swelling of the lips and tongue gives the tongue its typical blue appearance, though this sign is confined to a minority of the animals. Nasal signs may be prominent, with nasal discharge and stertorous respiration.

Some animals also develop foot lesions, beginning with coronitis, with consequent lameness. In sheep, this can lead to knee-walking. In cattle, constant changing of position of the feet gives bluetongue the nickname the dancing disease. Torsion of the neck (opisthotonos or torticollis) is observed in severely affected animals.

Not all animals develop signs, but all those that do lose condition rapidly, and the sickest die within a week. For affected animals that do not die, recovery is very slow, lasting several months.

The incubation period is 5–20 days, and all signs usually develop within a month. The mortality rate is normally low, but it is high in susceptible breeds of sheep. In Africa, local breeds of sheep may have no mortality, but in imported breeds, it may be up to 90%.

In cattle, goats, and wild ruminants, infection is usually asymptomatic despite high virus levels in blood. Red deer are an exception, and in them the disease may be as acute as in sheep.

MICROBIOLOGY

Bluetongue is caused by the pathogenic virus, Bluetongue virus (BTV), of the genus Orbivirus, of the Reoviridae family. Twenty-six serotypes are now recognised for this virus.

The virus particle consists of 10 strands of double-stranded RNA surrounded by two protein shells. Unlike other arboviruses, BTV lacks a lipid envelope. The particle has a diameter of 86 nm. The structure of the 70 nm core was determined in 1998 and was at the time the largest atomic structure to be solved.

The two outer capsid proteins, VP2 and VP5, mediate attachment and penetration of BTV into the target cell. The virus makes initial contact with the cell with VP2, triggering receptor-mediated endocytosis of the virus. The low pH within the endosome then triggers BTV's membrane penetration protein VP5 to undergo a conformational change that disrupts the endosomal membrane. Uncoating yields a transcriptionally active 470S core particle which is composed of two major proteins VP7 and VP3, and the three minor proteins VP1, VP4 and VP6 in addition to the dsRNA genome. There is no evidence that any trace of the outer capsid remains associated with these cores, as has been described for reovirus. The cores may be further uncoated to form 390S subcore particles that lack VP7, also in contrast to reovirus. Subviral particles are probably akin to cores derived in vitro from virions by physical or proteolytic treatments that remove the outer capsid and causes activation of the BTV transcriptase. In addition to the seven structural proteins, three non-structural (NS) proteins, NS1, NS2, NS3 (and a related NS3A) are synthesised in BTV-infected cells. Of these, NS3/NS3A is involved in the egress of the progeny virus. The two remaining non-structural proteins, NS1 and NS2, are produced at high levels in the cytoplasm and are believed to be involved in virus replication, assembly and morphogenesis.

EPIDEMIOLOGY

Bluetongue has been observed in Australia, the US, Africa, the Middle East, Asia, and Europe. An outline of the transmission cycle of BTV is illustrated in article Parasitic flies of domestic animals.

Its occurrence is seasonal in the affected Mediterranean countries, subsiding when temperatures drop and hard frosts kill the adult midge vectors. Viral survival and vector longevity is seen during milder winters. A significant contribution to the northward spread of bluetongue disease has been the ability of C. obsoletus and C.pulicaris to acquire and transmit the pathogen, both of which are spread widely throughout Europe. This is in contrast to the original C.imicola vector, which is limited to North Africa and the Mediterranean. The relatively recent novel vector has facilitated a far more rapid spread than the simple expansion of habitats north through global warming.

In August 2006, cases of bluetongue were found in the Netherlands, then Belgium, Germany, and Luxembourg.[14] In 2007, the first case of bluetongue in the Czech Republic was detected in one bull near Cheb at the Czech-German border. In September 2007, the UK reported its first ever suspected case of the disease, in a Highland cow on a rare-breeds farm near Ipswich, Suffolk. Since then, the virus has spread from cattle to sheep in Britain. By October 2007, bluetongue had become a serious threat in Scandinavia and Switzerland and the first outbreak in Denmark was reported. In autumn 2008, several cases were reported in the southern Swedish provinces of Småland, Halland, and Skåne, as well as in areas of the Netherlands bordering Germany, prompting veterinary authorities in Germany to intensify controls. Norway had its first finding in February 2009, when cows at two farms in Vest-Agder in the south of Norway showed an immune response to bluetongue. Norway have since been declared free of the disease in 2011.

Although the disease is not a threat to humans, the most vulnerable common domestic ruminants in the UK are cattle, goats, and especially, sheep.

Overwintering

A puzzling aspect of BTV is its survival between midge seasons in temperate regions. Adults of Culicoides are killed by cold winter temperatures, and BTV infections typically do not last for more than 60 days, which is not long enough for BTV to last until the next spring. It is believed that the virus somehow survives in overwintering midges or animals. Multiple mechanisms have been proposed. A few adult Culicoides midges infected with BTV may survive the mild winters of the temperate zone. Some midges may even move indoors to avoid the cold temperature of the winter. Additionally, BTV could cause a chronic or latent infection in some animals, providing another means for BTV to survive the winter. BTV can also be transmitted from mother to fetus. The outcome is abortion or stillbirth if fetal infection occurs early in gestation and survival if infection occurs late. However infection at an intermediate stage, before the fetal immune system is fully developed, may result in a chronic infection that lingers until the first months after birth of the lamb. Midges then spread the pathogen from the calves to other animals, starting a new season of infection.

Treatment and prevention

Prevention is effected via quarantine, inoculation with live modified virus vaccine, and control of the midge vector, including inspection of aircraft.

LIVESTOCK MANAGEMENT AND INSECT CONTROL

Vaccines

Protection by live attenuated vaccines (LAVs) are serotype specific. Multiserotype LAV cocktails can induce neutralizing antibodies against unincluded serotypes, and subsequent vaccinations with three different pentavalent LAV cocktails induce broad protection. These pentavalent cocktails contain 15 different serotypes in total: serotypes 1 through 14, as well as 19.

Immunization with any of the available vaccines, though, precludes later serological monitoring of affected cattle populations, a problem that could be resolved using next-generation subunit vaccines.

In January 2015, Indian researchers launched a vaccine named Raksha Blu that is designed to protect livestock against five strains of the bluetongue virus prevalent in the country.

History

Although bluetongue disease was already recognized in South Africa in the early 19th century, a comprehensive description of the disease was not published until the first decade of the 20th century. In 1906, Arnold Theiler showed that bluetongue was caused by a filterable agent. He also created the first bluetongue vaccine, which was developed from an attenuated BT V strain. For many decades, bluetongue was thought to be confined to Africa. The first confirmed outbreak outside of Africa occurred in Cyprus in 1943. Recently, a vessel owned by Khalifeh Livestock Trading and managed by Talia Shipping Line, both based in Lebanon, has been denied right to dock in Spain, as it has about 895 male calves suspected to be infected by bluetongue disease.

Historically, Bluetongue has been confined to tropical and subtropical areas (Central Africa, South East Asia) but climate change and trade patterns have seen increasing outbreaks in temperate regions (including Northern Europe) in recent years with outbreaks of up to 9 different serotypes occurring in Europe over the last 10 years. The most significant of which was the BTV-8 outbreak in Northern Europe in 2006-2008. Bluetongue (of any serotype) is currently absent in the UK but there is a significant risk of it being re-introduced with midges blown over the English Channel from continental Europe during European outbreaks. In particular, the current circulation of BTV-8 in southern France (as of autumn 2017) is a major risk to the UK.

Clinical Findings of Bluetongue in Ruminants

Bluetongue is characterised by changes to the mucous membranes of the mouth and nose, and the coronary band of the foot. Clinical signs are generally more severe in sheep but cattle can show signs of disease. A veterinary surgeon must be contacted by the farmer where large numbers of sheep or cattle present with lameness, high rectal temperatures, salivation, lacrimation and ocular and nasal discharges. Bluetongue is a notifiable disease in the UK.

The course of bluetongue in sheep can vary from peracute to chronic, with a mortality rate of 2%–90%. Peracute cases die within 7–9 days of infection, mostly as a result of severe pulmonary edema leading to dyspnea, frothing from the nostrils, and death by asphyxiation. In chronic cases, sheep may die 3–5 weeks after infection, mainly because of bacterial complications (especially pasteurellosis) and exhaustion. Animals with mild cases usually recover rapidly and completely. The major production losses include deaths, unthriftiness during prolonged convalescence, wool breaks, and reproductive losses.

In sheep, BTV causes vascular endothelial damage, resulting in changes to capillary permeability and subsequent intravascular coagulation, leading to edema, congestion, hemorrhage, inflammation, and necrosis. The clinical signs in sheep are typical. After an incubation period of 4–6 days, a fever of 40.5°–42°C (105°–107.5°F) develops. Affected animals are listless and reluctant to move. Clinical signs in young lambs are more apparent, and the mortality rate can be high (up to 30%). Approximately 2 days after onset of fever, additional clinical signs may be evident, such as edema of the lips, nose, face, submandibular area, eyelids, and sometimes ears; congestion of mouth, nose, nasal cavities, conjunctiva, and coronary bands; and lameness and depression. A serous nasal discharge is common, later becoming mucopurulent. The congestion of nose and nasal cavities produces a "sore muzzle" effect, the term used to describe the disease in sheep in the US.

Sheep eat less because of oral soreness and will hold food in their mouths to soften before chewing. They may champ to produce a frothy oral discharge at the corners of the lips. On close examination, small hemorrhages are visible on the mucous membranes of the nose and mouth. Ulceration develops where the teeth come in contact with lips and tongue, especially in the areas of most friction. Some affected sheep have severe swelling of the tongue, which may become cyanotic (hence the term "blue tongue") and even protrude from the mouth. Animals walk with difficulty as a result of inflammation of the hoof coronets. A purple-red color is obvious as a band at the junction of the skin and the hoof. Later in the course of disease, lameness or torticollis develops due to skeletal muscle damage. In most affected animals, abnormal wool growth results from dermatitis.

Clinical signs of bluetongue in cattle are rare but may be similar to those that occur in sheep. They are usually limited to fever, increased respiratory rate, lacrimation, salivation, stiffness, oral vesicles and ulcers, hyperesthesia, and a vesicular and ulcerative dermatitis. Susceptible cattle and sheep infected during pregnancy may abort or deliver malformed calves or lambs. The malformations include hydranencephaly or porencephaly, which results in ataxia and blindness at birth. White-tailed deer and pronghorn antelope develop severe hemorrhagic disease leading to sudden death.

In many areas of the world, BTV infection in sheep, and especially in other ruminants, is subclinical. Although clinical signs in cattle are rare and limited, exposure to BTV can influence international trade and movement of serologically positive animals.

Diagnosis of Bluetongue in Ruminants

- Clinical evaluation
- Postmortem examination
- Virus identification
- Serologic testing

The typical clinical signs of bluetongue enable a presumptive diagnosis, especially in areas where the disease is endemic.

Suspicion is confirmed by the presence of petechiae, ecchymoses, or hemorrhages in the wall of the base of the pulmonary artery and focal necrosis of the papillary muscle of the left ventricle. These highly characteristic lesions are usually obvious in severe clinical infections; in mild or convalescent cases, however, they may be barely visible. These lesions are often described as pathognomonic for bluetongue, but they also occur occasionally in other ovine diseases, such as heartwater, pulpy kidney disease, and Rift Valley fever. Hemorrhages and necrosis are usually found where mechanical abrasion damages fragile capillaries, such as on the buccal surface of the cheek opposite the molar teeth and the mucosa of the esophageal groove and omasal folds. Other necropsy findings include subcutaneous and intermuscular edema and hemorrhages, skeletal myonecrosis, myocardial and intestinal hemorrhages, hydrothorax, hydropericardium, pericarditis, and pneumonia.

Laboratory confirmation of bluetongue is based on virus isolation in embryonated chicken eggs or mammalian and insect cell cultures, or on identification of viral RNA by PCR assay. The identity of isolates may be confirmed by group-specific antigen-capture ELISA, group-specific PCR assay, immunofluorescence, immunoperoxidase technique, serotype-specific virus neutralization tests, serotype-specific PCR assay, or hybridization with complementary gene sequences of group- or serotype-specific genes. For virus isolation, blood (10–20 mL) is collected as early as possible from febrile animals into an anticoagulant such as heparin, sodium citrate, or ethylenediaminetetraacetic acid (EDTA) and transported at 4°C (39.2°F) to the laboratory. For long-term storage when refrigeration is not possible, blood is collected in oxalate-phenol-glycerin (OPG). Blood to be frozen should be collected in buffered lactose peptone and stored at or below -70°C (-94°F). Blood collected later during the viremic period should not be frozen because lysing of the RBCs on thawing releases the cell-associated virus, which may then be neutralized by early humoral antibody. The virus does not remain stable very long at -20°C (-4°F). In fatal cases, specimens of spleen, lymph nodes, or red bone marrow are collected and transported to the laboratory at 4°C (39.2°F) as soon as possible after death.

A serologic response in ruminants can be detected 7–14 days after infection by BTV and is generally lifelong after a field infection. Current recommended serologic techniques for detection of BTV antibody include agar gel immunodiffusion (AGID) and competitive ELISA. The latter is the test of choice and does not detect cross-reacting antibody to other orbiviruses, especially anti-EHDV (epizootic hemorrhagic disease virus) antibody. Various forms of the serum neutralization test, including plaque reduction, plaque inhibition, and microtiter neutralization, can be used to detect type-specific antibody.

Economic importance

Bluetongue virus infection has an enormous impact on sheep production in many countries on the African continent and elsewhere. Losses result primarily from mortality, reduced production during protracted convalescence including poor wool growth, and reduced reproductive performance including temporary ram infertility. Mortality rates can be high, with an average of 5% in the 2006 BTV-8 outbreak in the Netherlands, though in some flocks this was over 70%.

Sheep

The clinical signs of Bluetongue, which vary depending upon viral strain and sheep breed, follow an incubation period of four to 12 days. Usually, only a small percentage of sheep develop clinical signs, however deaths in some flocks can be as high as 70%. In extensively managed flocks, unexplained sudden deaths may be the first evidence of disease. Animals that survive the disease can lose condition with a reduction in meat and wool production.

Affected sheep have a fever (up to 42.0°C) and appear stiff and reluctant to move. They often adopt an arched back stance with the neck extended and the head held lowered. There is swelling of the face and ears, and also pulmonary oedema which may cause breathing difficulties. Erosions may appear on the lips progressing to ulcers. There is often profuse salivation, and a serous to mucopurulent nasal discharge. There may be reddening of the coronary band, and around the muzzle and mouth. The tongue may become swollen and lack of oxygen may make the tongue and mucous membranes appear blue (hence the name of the disease). Though this does not always occur. Bluetongue can also cause pregnant sheep to abort and infection during the breeding season may result in a large percentage of early embryonic losses with sheep returning to oestrus at irregular intervals. Foetal deformities similar to those seen with Schmallenberg virus can also sometimes occur.

- Key Points
- High rectal temperature
- Eye and nasal discharges
- Drooling as a result of ulcerations in the mouth
- Swelling of the mouth, head and neck.
- Lameness with inflammation at the junction of the skin and the coronary band
- Difficulty breathing
- Abortion

Differential diagnosis include foot and mouth disease. Orf may cause ulcertaions on the mucous membranes but is not usually a cause of high fever or mortality. Clostridial disease may cause sudden death and fever/oedema but this is usually sporadic and in unvaccinated sheep.

Cattle

Affected cattle are febrile (up to 40.0°C) and appear stiff due to swelling of the coronary band at the top of the hooves (coronary band) and are very reluctant to move. There is a serous to mucopurulent nasal discharge and erosions on the muzzle with sloughing of the mucosa. There is lacrimation but no obvious eye lesions. Mortality rates are usually much lower in cattle than in sheep.

Key Points

Fever up to (40.0°C)

Nasal discharge

Swelling of the head and neck

Conjunctivitis (runny eyes)

Swelling in, and ulceration of the mouth

Swollen teats

Saliva drooling out of the mouth

Abortion

Once again, the most important differential diagnosis is foot and mouth disease where there is profuse salivation, lameness and high rectal temperatures rapidly spreading to affect all cattle on the premises. The other differential diagnoses are infectious bovine rhinotracheitis (group or herd) and malignant catarrhal fever in individual cattle.

Diagnosis and treatment

Diagnosis is based upon clinical signs, virus detection via PCR and/or seroconversion to bluetongue virus. Treatment is limited to antibiotic therapy to control secondary bacterial infections.

Control and prevention

Control of bluetongue is very difficult because of the large number of potential hosts and virus serotypes. While control is aimed at keeping susceptible animals away from the vector this is not always practical. Control of the midges can be attempted with pour-on insecticides but this is expensive and does not achieve total freedom from the midge. Movement restrictions on affected animals may help with reducing spread to disease free regions but given how far the midges can blow restricting stock movements is of limited use in outbreaks.

The main prevention for BTV is vacciation. The BTV-8 vaccines available in the UK are killed vaccines. It is important to realize that here is no cross-protection between serotypes; vaccination against BTV-8 will not protect against other serotypes of BTV. The vaccines also do not act immediately with cattle requiring two doses of vaccine (and at least 6 weeks time from the first vaccination) to be protected. Importing unlicensed vaccines is not advised as there have been several outbreaks of disease due to the use of live virus vaccines imported illegally from South Africa into Northern Europe, these may also not be the correct serotype.

In many countries the timing of vaccination will depend upon local factors, in particular the occurrence of high-risk challenge periods from infected midges. Large scale vaccination efforts were important in controlling the BTV-8 outbreak in 2006-8, the virus however probably continued to circulate at low levels (in both domestic and wild ruminants) and re-emerged in 2016-17 when large numbers of niave animals were present again in European herds and flocks. Recent studies have shown that a vacciation effort of >95% of all susceptible animals for more than 5 years would be necessary to eliminate the virus altogether in Northern Europe. As this is unlikely to be achieved, monitoring of the current disease situation and choosing to vaccinate in advance of high risk periods is the only practical solution to preventing BTV at present.

15

FOOT-AND-MOUTH DISEASE

Foot-and-mouth disease (FMD) or hoof-and-mouth disease (HMD) is an infectious and sometimes fatal viral disease that affects cloven-hoofed animals, including domestic and wild bovids. The virus causes a high fever lasting two to six days, followed by blisters inside the mouth and on the feet that may rupture and cause lameness. FMD has very severe implications for animal farming, since it is highly infectious and can be spread by infected animals comparatively easily through contact with contaminated farming equipment, vehicles, clothing, and feed, and by domestic and wild predators. Its containment demands considerable efforts in vaccination, strict monitoring, trade restrictions, quarantines, and the culling of both infected and healthy (uninfected) animals.

Susceptible animals include cattle, water buffalo, sheep, goats, pigs, antelope, deer, and bison. It has also been known to infect hedgehogs and elephants; llamas and alpacas may develop mild symptoms, but are resistant to the disease and do not pass it on to others of the same species. In laboratory experiments, mice, rats, and chickens have been artificially infected, but they are not believed to contract the disease under natural conditions. Cattle, Asian and African buffalo, sheep, and goats can become carriers following an acute infection, meaning they are still infected with a small amount of virus but appear healthy. Animals can be carriers for up to 1–2 years and are considered very unlikely to infect other animals, although laboratory evidence suggests that transmission from carriers is possible.

Humans are only extremely rarely infected by foot-and-mouth disease virus (FMDV). (Humans, particularly young children, can be affected by hand, foot, and mouth disease (HFMDV), which is often confused for FMDV. Similarly, HFMDV is a viral infection belonging to the Picornaviridae family, but it is distinct from FMDV. HFMDV also affects cattle, sheep, and swine.)

The virus responsible for FMD is an aphthovirus, foot-and-mouth disease virus. Infection occurs when the virus particle is taken into a cell of the host. The cell is then forced to manufacture thousands of copies of the virus, and eventually bursts, releasing the new particles in the blood. The virus is genetically highly variable, which limits the effectiveness of vaccination. The disease was first documented in 1870.

SIGNS AND SYMPTOMS

The incubation period for FMD virus has a range between one and 12 days. The disease is characterized by high fever that declines rapidly after two to three days, blisters inside the mouth that lead to excessive secretion of stringy or foamy saliva and to drooling, and blisters on the feet that may rupture and cause lameness. Adult animals may suffer weight loss from which they do not recover for several months, as well as swelling in the testicles of mature males, and cows' milk production can decline significantly. Though most animals eventually recover from FMD, the disease can lead to myocarditis (inflammation of the heart muscle) and death, especially in newborn animals. Some infected ruminants remain asymptomatic carriers, but they nonetheless carry the virus and may be able to transmit it to others. Pigs cannot serve as asymptomatic carriers.

History

The cause of FMD was first shown to be viral in 1897 by Friedrich Loeffler. He passed the blood of an infected animal through a Chamberland filter and found the collected fluid could still cause the disease in healthy animals.

FMD occurs throughout much of the world, and while some countries have been free of FMD for some time, its wide host range and rapid spread represent cause for international concern. After World War II, the disease was widely distributed throughout the world. In 1996, endemic areas included Asia, Africa, and parts of South America; as of August 2007, Chile is disease-free, and Uruguay and Argentina have not had an outbreak since 2001. In May 2014, the FAO informed that Bolivia, Colombia, Ecuador and Peru were "just one step away" from eradication; North America and Australia have been free of FMD for many years. New Zealand has never had a case of foot-and-mouth disease. Most European countries have been recognized as disease-free, and countries belonging to the European Union have stopped FMD vaccination.

However, in 2001, a serious outbreak of FMD in Britain resulted in the slaughter of many animals, the postponing of the general election for a month, and the cancellation of many sporting events and leisure activities, such as the Isle of Man TT. Due to strict government policies on sale of livestock, disinfection of all persons leaving and entering farms, and the cancellation of large events likely to be attended by farmers, a potentially economically disastrous epizootic was avoided in the Republic of Ireland, with just one case recorded in Proleek, Co. Louth. As one result, the Animal Health Act 2002 was designed by Parliament to provide the regulators with more powers to deal with FMD.

In August 2007, FMD was found at two farms in Surrey, England. All livestock were culled and a quarantine erected over the area. Two other suspected outbreaks have occurred since, although these seem now not to be related to FMD. The only reported case in 2010 was a false alarm from GIS Alex Baker, as proven false by the Florida Farm and Agricultural Department, and quarantine/slaughter of cattle and pigs was confirmed from Miyazaki Prefecture in Japan in June after three cows tested positive. Some 270,000 cattle have been ordered slaughtered following the disease's outbreak.

Subclinical Infection

Subclinical (asymptomatic) infections can be classified as neoteric or persistent based on when they occur and whether the animal is infectious. Neoteric subclinical infections are acute infections, meaning they occur soon after an animal is exposed to the FMD virus (about 1 to 2 days) and last about 8 to 14 days. Acute infections are characterized by a high degree of replicating virus in the pharynx. In a neoteric subclinical infection, the virus remains in the pharynx and does not spread into the blood as it would in a clinical infection. Although animals with neoteric subclinical infections do not appear to have disease, they shed substantial amounts of virus in nasal secretions and saliva, so they are able to transmit the FMD virus to other animals. Neoteric subclinical infections often occur in vaccinated animals but can occur in unvaccinated animals as well.

Persistent subclinical infection (also referred to as a carrier state) occurs when an animal recovers from an acute infection but continues to have a small amount of replicating virus present in the pharynx. Cattle, buffalo, sheep, and goats can all become carriers, but pigs cannot. Animals can become carriers following acute infections with or without symptoms. Both vaccinated and unvaccinated animals can become carriers. Transmission of the FMD virus from carriers to susceptible animals is considered very unlikely under natural conditions and has not been conclusively demonstrated in field studies.

However, in an experiment where virus was collected from the pharynx of carrier cattle and inserted in the pharynx of susceptible cattle, the susceptible cattle became infected and developed characteristic blisters in the mouth and on the feet. This supports the theory that while the likelihood of a carrier spreading FMD is quite low, it is not impossible. It is not fully understood why ruminants but not pigs can become carriers or why some animals develop persistent infection while others do not. Both are areas of ongoing study.

Because vaccinated animals can become carriers, the waiting period to prove FMD-freedom is longer when vaccination rather than slaughter is used as an outbreak-control strategy. As a result, many FMD-free countries are resistant to emergency vaccination in case of in outbreak out of concern for the serious trade and economic implications of a prolonged period without FMD-free status.

Although the risk of transmission from an individual FMD carrier is considered to be very low, there are many carriers in FMD-endemic regions, possibly increasing the number of chances for carrier transmission to occur. Also, it can be difficult to determine if an asymptomatic infection is neoteric or persistent in the field, as both would be apparently healthy animals that test positive for the FMD virus. This fact complicates disease control, as the two types of subclinical infections have significantly different risks of spreading disease.

Cause

Of the seven serotypes of this virus, A, C, O, Asia 1, and SAT3 appear to be distinct lineages; SAT 1 and SAT 2 are unresolved clades. The mutation rate of the protein-encoding sequences of strains isolated between 1932 and 2007 has been estimated to be 1.46 × 10-3 substitutions/site/year, a rate similar to that of other RNA viruses. The most recent common ancestor appears to have evolved about 481 years ago (early 16th century).

This ancestor then diverged into two clades which have given rise to the extant circulating Euro-Asiatic and South African. SAT 1 diverged first 397 years ago, followed by sequential divergence of serotype SAT 2 (396 years ago), A (147 years ago), O (121 years ago), Asia 1 (89 years ago), C (86 years ago), and SAT 3 (83 years ago). Bayesian skyline plot reveals a population expansion in the early 20th century that is followed by a rapid decline in population size from the late 20th century to the present day. Within each serotype, there was no apparent periodic, geographic, or host species influence on the evolution of global FMD viruses. At least seven genotypes of serotype Asia 1 are known.

Transmission

The FMD virus can be transmitted in a number of ways, including close-contact, animal-to-animal spread, long-distance aerosol spread and fomites, or inanimate objects, typically fodder and motor vehicles. The clothes and skin of animal handlers such as farmers, standing water, and uncooked food scraps and feed supplements containing infected animal products can harbor the virus, as well. Cows can also catch FMD from the semen of infected bulls. Control measures include quarantine and destruction of both infected and healthy (uninfected) livestock, and export bans for meat and other animal products to countries not infected with the disease.

There is significant variation in both susceptibility to infection and ability to spread disease between different species, virus strains, and transmission routes. For example, cattle are far more vulnerable than pigs to infection with aerosolized virus, and infected pigs produce 30 times the amount of aerosolized virus compared to infected cattle and sheep. Also, pigs are particularly vulnerable to infection through the oral route. It has been demonstrated experimentally that FMD can be spread to pigs when they eat commercial feed products contaminated by the FMD virus. Also, the virus can remain active for extended periods of time in certain feed ingredients, especially soybean meal. Feed biosecurity practices have become an important area of study since a 2013 outbreak of Porcine Epidemic Diarrhea Virus (PEDV) in the US, thought to be introduced through contaminated feed.

Just as humans may spread the disease by carrying the virus on their clothes and bodies, animals that are not susceptible to the disease may still aid in spreading it. This was the case in Canada in 1952, when an outbreak flared up again after dogs had carried off bones from dead animals. Wolves are thought to play a similar role in the former Soviet Union.

Daniel Rossouw Kannemeyer (1843–1925) published a note in the Transactions of the South African Philosophical Society volume 8 part 1 in which he links saliva-covered locusts with the spread of the disease.

Transmission of the FMD virus is possible before an animal has apparent signs of disease, a factor that increases the risk that significant spread of the virus has occurred before an outbreak is detected. A 2011 experiment measured transmission timing in cattle infected with serotype O virus by exposing susceptible cattle in 24-hour increments. It estimated the infectious period of the infected cattle to be 1.7 days, but showed the cattle were only infectious for a few hours before they developed fevers or classic FMD lesions.

The authors also showed that the infectious period would have been estimated to be much higher (4.2 to 8.2 days) if detection of virus had been used as a substitute for infectiousness. A similar 2016 experiment using serotype A virus exposed susceptible pigs to infected pigs for 8 hour periods and found that pigs were able to spread disease for a full day before developing signs of disease. Analysis of this experimental data estimated the infectious period to be approximately 7 days. Again, the study showed that detection of virus was not an accurate substitution for infectiousness. An accurate understanding of the parameters of infectiousness is an important component of building epidemiological models which inform disease control strategies and policies.

Infecting humans

Humans can be infected with FMD through contact with infected animals, but this is extremely rare. Some cases were caused by laboratory accidents. Because the virus that causes FMD is sensitive to stomach acid, it cannot spread to humans via consumption of infected meat, except in the mouth before the meat is swallowed. In the UK, the last confirmed human case occurred in 1966, and only a few other cases have been recorded in countries of continental Europe, Africa, and South America. Symptoms of FMD in humans include malaise, fever, vomiting, red ulcerative lesions (surface-eroding damaged spots) of the oral tissues, and sometimes vesicular lesions (small blisters) of the skin. According to a newspaper report, FMD killed two children in England in 1884, supposedly due to infected milk.

Another viral disease with similar symptoms, hand, foot and mouth disease, occurs more frequently in humans, especially in young children; the cause, Coxsackie A virus, is different from the FMD virus. Coxsackie viruses belong to the Enteroviruses within the Picornaviridae.

Because FMD rarely infects humans, but spreads rapidly among animals, it is a much greater threat to the agriculture industry than to human health. Farmers around the world can lose enormous amounts of money during a foot-and-mouth epizootic, when large numbers of animals are destroyed, and revenues from milk and meat production go down.

Prevention

Like other RNA viruses, the FMD virus continually evolves and mutates, thus one of the difficulties in vaccinating against it is the huge variation between, and even within, serotypes. No cross-protection has been seen between serotypes (a vaccine for one serotype will not protect against any others) and in addition, two strains within a given serotype may have nucleotide sequences that differ by as much as 30% for a given gene. This means FMD vaccines must be highly specific to the strain involved. Vaccination only provides temporary immunity that lasts from months to years.

Currently, the World Organisation for Animal Health recognizes countries to be in one of three disease states with regard to FMD: FMD present with or without vaccination, FMD-free with vaccination, and FMD-free without vaccination. Countries designated FMD-free without vaccination have the greatest access to export markets, so many developed nations, including Canada, the United States, and the UK, work hard to maintain their current status. Some countries such as Brazil and Argentina, which have large beef-exporting industries, practise vaccination in some areas, but have other vaccination-free zones.

Reasons cited for restricting export from countries using FMD vaccines include, probably most importantly, routine blood tests relying on antibodies cannot distinguish between an infected and a vaccinated animal, which severely hampers screening of animals used in export products, risking a spread of FMD to importing countries. A widespread preventive vaccination would also conceal the existence of the virus in a country. From there, it could potentially spread to countries without vaccine programs. Lastly, an animal infected shortly after being vaccinated can harbor and spread FMD without showing symptoms itself, hindering containment and culling of sick animals as a remedy.

Many early vaccines used dead samples of the FMD virus to inoculate animals, but those early vaccines sometimes caused real outbreaks. In the 1970s, scientists discovered that a vaccine could be made using only a single key protein from the virus. The task was to produce enough quantities of the protein to be used in the vaccination. On June 18, 1981, the US government announced the creation of a vaccine targeted against FMD, the world's first genetically engineered vaccine.

The North American FMD Vaccine Bank is housed at the United States Department of Agriculture's Foreign Animal Disease Diagnostic Laboratory at Plum Island Animal Disease Center. The center, located 1.5 mi (2.4 km) off the coast of Long Island, New York, is the only place in the United States where scientists can conduct research and diagnostic work on highly contagious animal diseases such as FMD. Because of this limitation, US companies working on FMD usually use facilities in other countries where such diseases are endemic.

United States 1870–1929

The US has had nine FMD outbreaks since it was first recognized on the northeastern coast in 1870; the most devastating happened in 1914. It originated from Michigan, but its entry into the stockyards in Chicago turned it into an epizootic. About 3,500 livestock herds were infected across the US, totaling over 170,000 cattle, sheep, and swine. The eradication came at a cost of US$4.5 million (equivalent to $122 million in 2021).

A 1924 outbreak in California resulted not only in the slaughter of 109,000 farm animals, but also 22,000 deer.

The US had its latest FMD outbreak in Montebello, California, in 1929. This outbreak originated in hogs that had eaten infected meat scraps from a tourist steamship that had stocked meat in Argentina. Over 3,600 animals were slaughtered and the disease was contained in less than a month.

Mexico–U.S. border 1947

On December 26, 1946, the United States and Mexico jointly declared that FMD had been found in Mexico. Initially, proposals from Texans were for an animal-proof wall, to prevent animals from crossing the border and spreading the disease, but the two countries eventually managed to cooperate in a bilateral effort and eradicated the disease without building a wall. To prevent tension between ranchers and the veterinarians, public broadcasts over the radio and with speakers on trucks were used to inform Mexican ranchers of why the U.S. veterinarians were working on their livestock. Any rancher who lost cattle due to it being put down by the vets would receive financial compensation.

However, the tension remained and unfortunately resulted in clashes between locals and the military protected U.S. veterinarians. These teams of veterinarians worked from outside the infection zone of the disease and worked their way to the heart of the epidemic. Over 60,000,000 injections were administered to livestock by the end of 1950. It was a sign that U.S.–Mexico relations were improving in the post-World War II era.

United Kingdom 1967

In October 1967, a farmer in Shropshire reported a lame sow, which was later diagnosed with FMD. The source was believed to be remains of legally imported infected lamb from Argentina and Chile. The virus spread, and in total, 442,000 animals were slaughtered and the outbreak had an estimated cost of £370 million (equivalent to £7 billion in 2021).

Taiwan 1997

Taiwan had previous epidemics of FMD in 1913–14 and 1924–29, but had since been spared, and considered itself free of FMD as late as in the 1990s. On the 19th of March 1997, a sow at a farm in Hsinchu, Taiwan, was diagnosed with a strain of FMD that only infects swine. Mortality was high, nearing 100% in the infected herd. The cause of the epidemic was not determined, but the farm was near a port city known for its pig-smuggling industry and illegal slaughterhouses. Smuggled swine or contaminated meat are thus likely sources of the disease.

The disease spread rapidly among swine herds in Taiwan, with 200–300 new farms being infected daily. Causes for this include the high swine density in the area, with up to 6,500 hogs per square mile, feeding of pigs with untreated garbage, and the farms' proximity to slaughterhouses. Other systemic issues, such as lack of laboratory facilities, slow response, and initial lack of a vaccination program, contributed.

A complicating factor is the endemic spread of swine vesicular disease (SVD) in Taiwan. The symptoms are indistinguishable from FMD, which may have led to previous misdiagnosing of FMD as SVD. Laboratory analysis was seldom used for diagnosis, and FMD may thus have gone unnoticed for some time.

The swine depopulation was a massive undertaking, with the military contributing substantial manpower. At peak capacity, 200,000 hogs per day were disposed of, mainly by electrocution. Carcasses were disposed of by burning and burial, but burning was avoided in water resource-protection areas. In April, industrial incinerators were running around the clock to dispose of the carcasses.

Initially, 40,000 combined vaccine doses for the strains O-1, A-24, and Asia-1 were available and administered to zoo animals and valuable breeding hogs. At the end of March, half a million new doses for O-1 and Asia-1 were made available. On the May 3rd, 13 million doses of O-1 vaccine arrived, and both the March and May shipments were distributed free of charge. With a danger of vaccination crews spreading the disease, only trained farmers were allowed to administer the vaccine under veterinary supervision.

Taiwan had previously been the major exporter of pork to Japan, and among the top 15 pork producers in the world in 1996. During the outbreak, over 3.8 million swine were destroyed at a cost of US$6.9 billion (equivalent to $11.6 billion in 2021). The Taiwanese pig industry was devastated as a result, and the export market was in ruins. In 2007, Taiwan was considered free of FMD, but was still conducting a vaccination program, which restricts the export of meat from Taiwan.

United Kingdom 2001

The epidemic of FMD in the United Kingdom in the spring and summer of 2001 was caused by the "Type O pan Asia" strain of the disease. This episode resulted in more than 2,000 cases of the disease in farms throughout the British countryside. More than six million sheep and cattle were killed in an eventually successful attempt to halt the disease. The county of Cumbria was the most seriously affected area of the country, with 843 cases. By the time the disease was halted in October 2001, the crisis was estimated to have cost Britain £8 billion (equivalent to £14 billion in 2021) to the agricultural and support industries, and to the outdoor industry. What made this outbreak so serious was the amount of time between infection being present at the first outbreak locus, and when countermeasures were put into operation against the disease, such as transport bans and detergent washing of both vehicles and personnel entering livestock areas. The epidemic was probably caused by pigs that had been fed infected rubbish that had not been properly heat-sterilized. Further, the rubbish is believed to have contained remains of infected meat that had been illegally imported to Britain.

China 2005

In April 2005, an Asia-1 strain of FMD appeared in the eastern provinces of Shandong and Jiangsu. During April and May, it spread to suburban Beijing, the northern province of Hebei, and the Xinjiang autonomous region in northwest China. On 13 May, China reported the FMD outbreak to the World Health Organization and the OIE. This was the first time China has publicly admitted to having FMD. China is still reporting FMD outbreaks. In 2007, reports filed with the OIE documented new or ongoing outbreaks in the provinces of Gansu, Qinghai and Xinjiang. This included reports of domestic yak showing signs of infection.

FMD is endemic in pastoral regions of China from Heilongjiang Province in the northeast to Sichuan Province and the Tibetan Autonomous region in the southwest. Chinese domestic media reports often use a euphemism "Disease Number Five" (??? wuhàobìng) rather than FMD in reports because of the sensitivity of the FMD issue. In March 2010, Southern Rural News (Nanfang Nongcunbao), in an article "Breaking the Hoof and Mouth Disease Taboo", noted that FMD has long been covered up in China by referring to it that way. FMD is also called canker (??, literally "mouth ulcers" kouchuang) or hoof jaundice (?? tíhuáng) in China, so information on FMD in China can be found online using those words as search terms.

One can find online many provincial orders and regulations on FMD control antedating China's acknowledgment that the disease existed in China, for example Guangxi Zhuang Autonomous Region 1991 regulation on preventing the spread of Disease No.5.

United Kingdom 2007

An infection of FMD in the United Kingdom was confirmed by the Department for Environment, Food and Rural Affairs, on 3 August 2007, on farmland located in Normandy, Surrey. All livestock in the vicinity were culled on 4 August. A nationwide ban on the movement of cattle and pigs was imposed, with a 3-km (1.9-mi) protection zone placed around the outbreak sites and the nearby virus research and vaccine production establishments, together with a 10-km (6.2-mi) increased surveillance zone.

On 4 August, the strain of the virus was identified as a "01 BFS67-like" virus, one linked to vaccines and not normally found in animals, and isolated in the 1967 outbreak. The same strain was used at the nearby Institute for Animal Health and Merial Animal Health Ltd at Pirbright, 2.5 miles (4.0 km) away, which is an American/French-owned research facility, and was identified as the likely source of infection.

On 12 September, a new outbreak of the disease was confirmed in Egham, Surrey, 19 km (12 mi) from the original outbreak, with a second case being confirmed on a nearby farm on 14 September.

These outbreaks caused a cull of all at-risk animals in the area surrounding Egham, including two farms near the famous four-star hotel Great Fosters. These outbreaks also caused the closure of Windsor Great Park due to the park containing deer; the park remained closed for three months. On 19 September 2007, a suspected case of FMD was found in Solihull, where a temporary control zone was set up by Defra.

Japan and Korea 2010–2011

In April 2010, a report of three incursions of FMD in Japan and South Korea led the United Nations Food and Agriculture Organization (FAO) to issue a call for increased global surveillance. Japan veterinary authorities confirmed an outbreak of type O FMD virus, currently more common in Asian countries where FMD is endemic.

South Korea was hit by the rarer type A FMD in January, and then the type O infection in April. The most serious case of foot-and-mouth outbreak in South Korea's history started in November 2010 in pig farms in Andong city of Gyeongsangbuk-do, and has since spread in the country rapidly. More than 100 cases of the disease have been confirmed in the country so far, and in January 2011, South Korean officials started a mass cull of approximately 12%, or around three million in total, of the entire domestic pig population, and 107,000 of three million cattle of the country to halt the outbreak. According to the report based on complete 1D gene sequences, Korean serotype A virus was linked with those from Laos. Korean serotype O viruses were divided into three clades and were closely related to isolates from Japan, Thailand, the UK, France, Ireland, South Africa, and Singapore, as well as Laos.

On 10 February 2011, North Korea reported an outbreak affecting pigs in the region around Pyongyang, by then ongoing since at least December 2010. Efforts to control the outbreak were hampered by illicit sales of infected meat.

Indonesia in 2022

After eradication in 1986, FMD was detected in Indonesia in May 2022. The Australian government has offered its assistance but remains unconcerned, considering the risk to the country's biosecurity to be low. The Department of Agriculture (DAWE) is the responsible body and has been monitoring the situation. DAWE has determined there is only a low risk and has stockpiled vaccines since 2004 anyhow.

SOCIETY AND CULTURE

Economic and ethical issues

Epidemics of FMD have resulted in the slaughter of millions of both infected and healthy animals, despite this being a frequently nonfatal disease for adult animals (2–5% mortality), though young animals can have a high mortality. The 1997 Taiwan outbreak that affected only pigs also showed a high mortality for adults. The destruction of animals is primarily to halt further spread, as growth and milk production may be permanently affected, even in animals that have recovered.

Due to international efforts to eradicate the disease, infection would also lead to trade bans being imposed on affected countries. Critics of current policies to cull herds argue that the financial imperative needs to be balanced against the killing of many, mostly healthy animals, especially when a significant proportion of infected animals, most notably those producing milk, would recover from infection and live normal lives, albeit with reduced milk production. On the ethical side, one must also consider that FMD is a painful disease for the affected animals.

The vesicles and blisters are painful in themselves, and restrict both eating and movement. Through ruptured blisters, the animal is at risk from secondary bacterial infections and, in some cases, permanent disability.

BIBLIOGRAPHY

- Hoernle, A. F. Rudolf (August Friedrich Rudolf) (1907). Studies in the medicine of ancient India. Gerstein – University of Toronto. Oxford : At the Clarendon Press.
- Byrne, Joseph Patrick (2012). Encyclopedia of the Black Death. ABC-CLIO. p. 29. ISBN 978-1598842531.
- Reid, Megan H. (2013). Law and Piety in Medieval Islam. Cambridge University Press. pp. 106, 114, 189–90. ISBN 978-1107067110.
- Majeed, Azeem (22 December 2005). "How Islam changed medicine". BMJ. 331 (7531): 1486–87. doi:10.1136/bmj.331.7531.1486. ISSN 0959-8138. PMC 1322233. PMID 16373721.
- Beretta M (2003). "The revival of Lucretian atomism and contagious diseases during the renaissance". Medicina Nei Secoli. 15 (2): 129–54. PMID 15309812.
- Moorhead Robert (November 2002). "William Budd and typhoid fever". J R Soc Med. 95 (11): 561–64. doi:10.1177/014107680209501115. PMC 1279260. PMID 12411628.
- Sawyer RT (1986). "Leech Biology and Behaviour". Feeding, Biology. Ecology and Systematic (PDF). Vol. II. Oxford: Clarendon Press. Retrieved 18 August 2020 – via Biopharm Leeches.
- Hazen TC, Fliermans CB, Hirsch RP, Esch GW (November 1978). "Prevalence and distribution of Aeromonas hydrophila in the United States". Applied and Environmental Microbiology. 36 (5): 731–8. Bibcode:1978ApEnM..36..731H. doi:10.1128/aem.36.5.731-738.1978. PMC 243130. PMID 31839.
- Agger WA, McCormick JD, Gurwith MJ (June 1985). "Clinical and microbiological features of Aeromonas hydrophila-associated diarrhea". Journal of Clinical Microbiology. 21 (6): 909–13. doi:10.1128/jcm.21.6.909-

913.1985. PMC 271816. PMID 4008621.

- Tan WS, Yin WF, Chan KG (January 2015). "Insights into the Quorum-Sensing Activity in Aeromonas hydrophila Strain M013 as Revealed by Whole-Genome Sequencing". Genome Announcements. 3 (1): 1–2. doi:10.1128/genomeA.01372-14. PMC 4293626. PMID 25555739.
- Snower DP, Ruef C, Kuritza AP, Edberg SC (June 1989). "Aeromonas hydrophila infection associated with the use of medicinal leeches". Journal of Clinical Microbiology. 27 (6): 1421–2. doi:10.1128/jcm.27.6.1421-1422.1989. PMC 267578. PMID 2666448.
- "Cellulitis". The Lecturio Medical Concept Library. Retrieved 7 July 2021.
- Nettle D (2009). "Ecological influences on human behavioural diversity: a review of recent findings". Trends Ecol. Evol. 24 (11): 618–24. doi:10.1016/j.tree.2009.05.013. PMID 19683831.
- Molnar, R. E., 2001, "Theropod paleopathology: a literature survey": In: Mesozoic Vertebrate Life, edited by Tanke, D. H., and Carpenter, K., Indiana University Press, pp. 337–63.
- Caspermeyer, Joe (23 September 2007). "Space flight shown to alter ability of bacteria to cause disease". Arizona State University. Retrieved 14 September 2017.
- Kim W, et al. (April 29, 2013). "Spaceflight Promotes Biofilm Formation by Pseudomonas aeruginosa". PLOS ONE. 8 (4): e6237. Bibcode:2013PLoSO...862437K. doi:10.1371/journal.pone.0062437. PMC 3639165. PMID 23658630.
- Dvorsky, George (13 September 2017). "Alarming Study Indicates Why Certain Bacteria Are More Resistant to Drugs in Space". Gizmodo. Retrieved 14 September 2017.

INDEX

www.ingramcontent.com/pod-product-compliance
Ingram Content Group UK Ltd.
Pitfield, Milton Keynes, MK11 3LW, UK
UKHW021328190726
13845UKWH00004B/70

9 789349 964815